ISBN 978-0-331-83481-9
PIBN 11045768

TREES AND SHRUBS

ILLUSTRATIONS OF

NEW OR LITTLE KNOWN LIGNEOUS PLANTS

PREPARED CHIEFLY FROM MATERIAL AT
THE ARNOLD ARBORETUM OF
HARVARD UNIVERSITY

AND EDITED BY

CHARLES SPRAGUE SARGENT

Director of the Arnold Arboretum; Author of
The Silva of North America

VOLUME I

BOSTON AND NEW YORK

Part I., Plates I.-XXV., issued November 26, 1902.
Part II., Plates XXVI.-L., issued May 13, 1903.
Part III., Plates LI.-LXXV., issued November 14, 1903.
Part IV., Plates LXXVI.-C., issued April 8, 1905.

INTRODUCTION

This work consists of a series of plates accompanied by brief descriptions of new or little known trees and shrubs of the northern hemisphere which may be expected to flourish in the gardens of the United States and Europe, and those of special economic value or scientific interest. The material which serves as a basis for the work has been derived largely from the living collections and from the herbarium of the Arnold Arboretum. In the preparation of this work I have had the assistance of a number of specialists who have signed the articles contributed by them. The plates are reproductions of original drawings made by Mr. C. E. Faxon. The first volume contains one hundred plates and the descriptions of two genera, forty-eight species, and a number of varieties new to science, and of a number of plants recently introduced into gardens.

C. S. Sargent.

Arnold Arboretum,
February, 1905.

TABLE OF CONTENTS.

TREES AND SHRUBS.

JUGLANS MEXICANA, S. WATS.

JUGLANS MEXICANA, S. Watson, *Proc. Am. Acad.* xxvi. 152 (1891).

Leaves eleven to fifteen-foliolate, from 3 to 4 decimetres long, with slender tomentose ultimately nearly glabrous rachises; leaflets ovate, acuminate, often slightly falcate, very unequal at the base, finely serrate, with minute incurved teeth tipped with dark glands, and slightly thickened and revolute on the margins, scabrate but ultimately nearly glabrous on the upper surface, coated below in early spring with pale tomentum, at maturity soft-pubescent below particularly along the midribs and in the axils of the veins, sessile or short-petiolulate, from 10 to 12.5 centimetres long and 4 centimetres wide, the terminal leaflet raised on a slender stalk often nearly 2.5 centimetres in length. Staminate flowers opening when the leaves are about a third grown, pedicellate in thick catkins from 7.5 to 12.5 centimetres in length; perianth six-lobed, the lobes rounded above, pubescent toward the base, its bract villose at the apex; stamens from eighteen to twenty-five in several ranks; anthers surmounted by short bifid connectives. Pistillate flowers often solitary, ovate, 4 millimetres long, their bracts and bractlets coated with hoary tomentum; stigmas clavate, spreading, bright red, 8 millimetres in length. Fruit oblong, short-pointed, covered with small orange-brown lenticels, puberulous especially toward the ends, from 4 to 6 centimetres long and from 2.5 to 5 centimetres broad; exocarp spongy, one third of an inch in thickness. Nut oblong to oblong-obovate, obtuse or slightly apiculate at the apex, bluntly ridged along the dorsal sutures, deeply sulcate, with longitudinal simple or forked grooves, reddish brown, the shell thick, lacunose, with large cavities. Lobes of the seed prominently keeled on the back, slightly concave and keeled on the inner face.

A tree, often 20 metres in height, with a trunk from 1 to 1.5 metres in diameter covered with thick deeply furrowed nearly black bark, stout wide-spreading branches forming an open rather irregular head, and comparatively slender branchlets coated, when they first appear, with short close pale tomentum, becoming pubescent during their second season and puberulous during their third year.

Mountain valleys, Monterey, Mexico, at elevations of 800 metres above the sea, *C. S. Sargent*, March, 1887; hills at San José Pass, San Luis Potosí, *C. G. Pringle* (No. 3322), October, 1890; near Monterey, *Canby, Trelease, and Sargent*, March 19, 1900, in flower.

This handsome and interesting tree, which is very common in cañons of the mountains near Monterey, has the habit, bark, and the foliage of *Juglans nigra*, Linnæus; the pubescence which clothes the leaves and young branchlets, however, resembles that of *Juglans cinerea*, Linnæus. The sulcate nut shows its relationship with the Texas *Juglans rupestris*, Engelmann, although much larger than the nuts of that species, which are destitute of the thick ridges on the dorsal sutures. The bifid connective of the anthers resembles that of *Juglans Californica*, Watson. From the little known *Juglans mollis*, Hemsley, it differs in the form and texture of the leaflets and in their venation and pubescent covering.

C. S. S.

EXPLANATION OF THE PLATE.

PLATE I. JUGLANS MEXICANA.

1. A flowering branch, natural size.
2. A staminate flower, enlarged.
3. An anther, enlarged.
4. A fruiting branch, natural size.
5. A fruit, part of the husk removed, natural size.
6. Cross section of a nut, natural size.

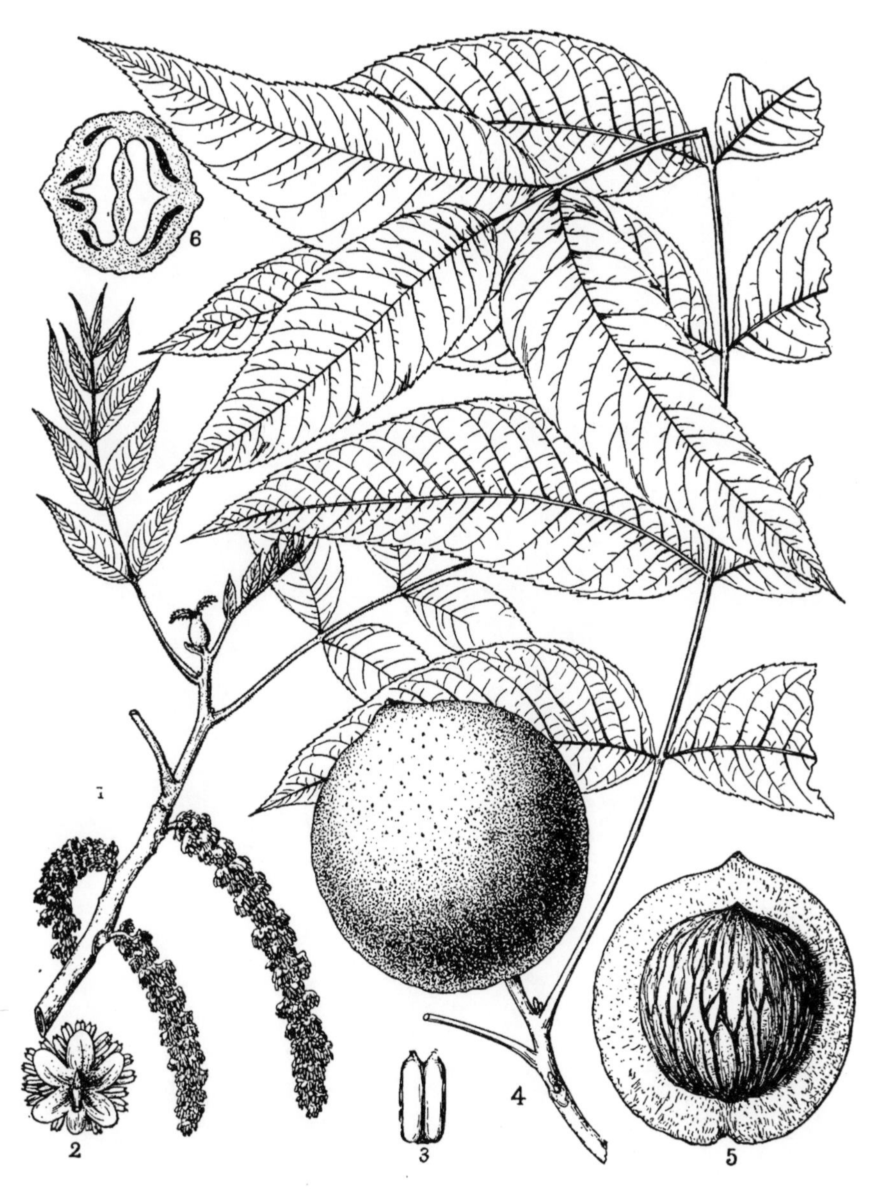

C. E. Faxon del.

JUGLANS MEXICANA, S. Wats.

CRATÆGUS DUROBRIVENSIS, Sarg.

Cratægus Durobrivensis, *n. sp.*

Leaves ovate, acute, slightly divided above the middle into two or three pairs of pointed lobes, sharply serrate except at the base, with slender spreading or incurved teeth tipped with minute red glands, membranaceous, at first hirsute, with a few scattered white hairs on the upper side of the slender midribs, soon entirely glabrous, dark yellow-green on the upper surface, paler on the lower surface, the primary veins slender and slightly impressed above, from 5 to 6.5 centimetres long and from 3 to 4 centimetres wide; petioles slender, deeply grooved, glandular, from 1.3 to 1.9 centimetres in length, dark red below the middle; leaves on vigorous leading shoots often ovate, rounded at the broad base, more deeply divided into broader lobes, more or less decurrent on the stouter petioles, and from 6.5 to 7.5 centimetres long and often 6.5 centimetres wide. Flowers from 2 to 2.5 centimetres in diameter, in many-flowered compact glabrous compound corymbs; bracts and bractlets linear, conspicuously glandular, caducous; calyx-cavity broad and shallow, the lobes lanceolate, glandular-serrate, villose on the upper surface, with scattered pale hairs, reflexed after anthesis; stamens twenty; filaments slender, elongated; anthers rose color; styles five, surrounded at the base by broad tufts of thick white tomentum. Fruit globose, about 1.6 centimetres in diameter, slightly depressed at the insertion of the slender peduncle, dark crimson, lustrous, marked by few small pale dots; calyx prominent, persistent, with a deep broad cavity, and closely appressed linear-lanceolate lobes abruptly narrowed from broad bases, bright red on their upper surface, irregularly glandular serrate above the middle; flesh thick, pale yellow, sweet, dry, and mealy; nutlets five, rounded and obscurely grooved or slightly ridged on the back, dark-colored, 6 millimetres long.

A broad shrub, from 3 to 6 metres in height, with numerous stout intricately branched stems covered with close smooth bark ashy gray near the ground and olive-green above, and slender branchlets dark green and glabrous when they first appear, soon becoming dark chestnut-brown and lustrous, dull ashy gray in their second season, and furnished with stout straight or slightly curved dark chestnut-brown spines from 3.8 to 5 centimetres in length and on the large stems often compound and from 7.5 to 10 centimetres long. Flowers at the end of May when the leaves are about half-grown. Fruit ripens in October and remains on the branches without loss of color until the new year and long after the leaves have fallen.

Steep rocky banks of the Genesee River, Rochester, New York, *John Dunbar*, May, 1900; banks of the Niagara River, New York, *C. S. Sargent*, September, 1901; Buffalo, New York, *John Dunbar*, September, 1901.

Cratægus Durobrivensis belongs to a small section of the genus of which *Cratægus dilatata*, Sargent, may be considered the type, and which may be distinguished by large flowers with twenty stamens and red anthers, and by large brightly colored fruit crowned with the much enlarged and conspicuous calyx, which is usually bright red on the upper surface of the base of the lobes. *Cratægus Durobrivensis* differs from the other known members of this group in the absence of pubescence and in the persistence of the fruit, which does not fall until midwinter. This peculiarity and the large showy flowers make it one of the most ornamental of the Thorns of the northern United States.

C. S. S.

EXPLANATION OF THE PLATE.

PLATE II. CRATÆGUS DUROBRIVENSIS.

1. Part of a flowering branch, natural size.
2. Vertical section of a flower, enlarged.
3. A calyx-lobe, enlarged.
4. A fruiting branch, natural size.
5. Cross section of a fruit, enlarged.
6. Vertical section of a fruit, enlarged.
7. A nutlet, side view, enlarged.
8. A nutlet, rear view, enlarged.

C. E. Faxon del.

CRATÆGUS DUROBRIVENSIS, Sarg.

CRATÆGUS COLEÆ, SARG.

CRATÆGUS COLEÆ, *n. sp.*

Leaves ovate to oval, acute, broadly cuneate at the entire base, irregularly and often doubly serrate above, with small incurved teeth tipped with minute dark glands, and divided into numerous short acute lateral lobes; glabrous as they unfold with the exception of a few scattered caducous pale hairs on the upper surface, at maturity membranaceous, dark green and lustrous above, pale and somewhat glaucous below, from 4.5 to 5 centimetres long and from 3 to 3.5 centimetres wide; petioles slender, nearly terete, from 1.2 to 1.5 centimetres long; stipules linear, coarsely glandular-serrate, caducous; leaves on vigorous shoots often full and rounded at the base and more coarsely serrate and more deeply lobed than the leaves of fertile branchlets. Flowers from 18 to 20 millimetres in diameter on elongated slender pedicels, in broad compound many-flowered corymbs; bracts and bractlets linear, conspicuously glandular-serrate, caducous; calyx-tube broadly obconic, the lobes gradually narrowed from broad bases, acuminate, glandular-serrate, with small bright red stipitate glands, villose on the inner surface particularly toward the base, reflexed after anthesis; stamens eighteen or twenty; anthers large, tinged with pink; styles two to five, usually two to four, surrounded at the base by a broad ring of pale tomentum. Fruit pendant on slender pedicels, in many-fruited clusters, subglobose, often somewhat broader than long, bright orange-red marked by small pale dots, from 8 to 10 millimetres in diameter; calyx enlarged and prominent, with a broad deep tube and elongated reflexed anther-lobes sometimes deciduous before the fruit falls; nutlets two to five, broad, obtuse at the ends, 7 to 8 millimetres long, prominently ridged on the back, with a high broad ridge, deeply penetrated on each of the inner faces by a long narrow longitudinal depression.

A broad tree-like shrub, with numerous stout spreading stems covered with close dark bark and from 4 to 5 metres tall, and slender nearly straight or slightly zigzag glabrous branchlets marked by oblong pale lenticels, light yellow-green when they first appear, bright red-brown and very lustrous during their first season and dark gray-brown the following year, and armed with numerous nearly straight stout lustrous spines often pointed toward the base of the branch and from 2.5 to 4.5 centimetres long. Flowers during the last week of May. Fruit ripens at the end of September.

Hillsides in rich moist soil near Grand Rapids, Michigan, *Miss E. J. Cole*, May and September, 1901.

Although it differs from *Cratægus tomentosa*, Linnæus, in the absence of pubescence and in the size and shape of the fruit, and from all the species related to *Cratægus macracantha*, Lindley, in its thin leaves with their slender veins, the longitudinal grooves on the inner face of the nutlets indicate the relationship of *Cratægus Coleæ* with the Tomentosa group, in which it may form the type of an interesting and distinct subsection.

It is a pleasure to associate with this handsome shrub the name of its discoverer, Miss Emma J. Cole of Grand Rapids, Michigan, the author of *The Grand Rapids Flora*, and a careful and industrious student of the plants of central Michigan, where she has made a number of other important discoveries. C. S. S.

EXPLANATION OF THE PLATE.

PLATE IV. CRATÆGUS COLEÆ.

1. A flowering branch, natural size.
2. Vertical section of a flower, enlarged.
3. A calyx-lobe, enlarged.
4. A fruiting branch, natural size.
5. Vertical section of a fruit, enlarged.
6. Cross section of a fruit, enlarged.
7. A nutlet, side view, enlarged.
8. A nutlet, rear view, enlarged.

C. E. Faxon del.

CRATÆGUS COLEÆ, Sarg.

CRATÆGUS MALOIDES, SARG.

CRATÆGUS MALOIDES, *n. sp.*

Leaves oblong-obovate to oval, gradually narrowed from near the middle and acute or full and rounded at the apex, cuneate and entire below, finely and usually doubly serrate above, with broad straight or incurved glandular teeth; when they unfold bronze-red and slightly hairy on the upper side of the midribs, and furnished on the under surface with large tufts of pale tomentum in the axils of the veins, at maturity coriaceous, glabrous, dark green and lustrous above, pale yellow-green below, from 3 to 3.5 centimetres long and from 1 to 1.5 centimetres wide, with slender midribs and few thin primary veins running obliquely toward the apex of the leaf and only slightly impressed above; petioles slender, more or less wing-margined above, glandular, villose at first but soon glabrous, often red, about 1 centimetre long; stipules linear, minute, caducous; leaves on vigorous shoots frequently broadly oval, slightly divided into short lateral lobes and from 3.5 to 4 centimetres long and broad, their stipules foliaceous, coarsely glandular-serrate. Flowers 2 centimetres in diameter on slender pedicels, in three-flowered glabrous corymbs subtended by one or two solitary flowers from the axils of the upper leaves; bracts and bractlets linear, minute, finely glandular-serrate, bright red; calyx-tube broadly obconic, glabrous, the lobes linear-lanceolate, elongated, bright red, reflexed after anthesis; stamens fifteen to twenty, anthers large, dark red; styles three to five. Fruit usually solitary or in two or three-fruited clusters on stout red pedicels from 1.2 to 2.5 centimetres long, short-oblong to obovate, truncate at the broad apex, gradually narrowed and full and rounded at the base, with a deep depression at the insertion of the stalk, bright cherry-red and very lustrous, from 1.5 to 1.7 centimetres long and about 1.2 centimetres wide; flesh thick, bright yellow, very juicy, subacid; calyx-cavity narrow and shallow, the lobes gradually narrowed from broad bases, entire, wide-spreading; nutlets three to five, comparatively small, rounded and obscurely ridged on the back, about 7 millimetres long.

A tall shrub, with numerous wide-spreading slender stems and thin slightly zigzag branchlets marked by small white oblong lenticels, dark red and covered with scattered pale hairs when they first appear, soon glabrous, dark mahogany-red and lustrous during their first summer, dull ashy gray during their second year, and armed with few straight chestnut-brown spines from 2 to 2.5 centimetres in length. Flowers during the first week of March. Fruit ripens early in June and falls gradually until the beginning of July.

Borders of wet clay prairies at Haw Creek, six miles east of Seville, Volusia County, Florida, *A. H. Curtiss* (No. 6679), July, 1900, March and June, 1901.

The large flowers in few-flowered corymbs with their numerous stamens and large dark red anthers and the large early ripening fruit of this very distinct species point to its relationship with *Cratægus æstivalis* of Torrey & Gray. The fruit, which is very beautiful and more juicy than that of any other Cratægus I have seen, is distinctly subacid and of a true apple flavor.

C. S. S.

EXPLANATION OF THE PLATE.

PLATE V. CRATÆGUS MALOIDES.

1. A flowering branch, natural size.
2. Vertical section of a flower, enlarged.
3. A calyx-lobe, enlarged.
4. A fruiting branch, natural size.
5. Vertical section of a fruit, natural size.
6. Cross section of a fruit showing the nutlets, natural size.
7. A nutlet, side view, enlarged.

C. E. Faxon del.

CRATÆGUS MALOIDES, Sarg.

CRATÆGUS LUCULENTA, Sarg.

Cratægus luculenta, *n. sp.*

Leaves oblong-obovate or rhomboidal, acute or rounded at the apex, crenulate-serrate above the middle, with glandular teeth, gradually narrowed, cuneate and entire below, sometimes divided into short lateral lobes; as they unfold slightly villose above, with short caducous hairs, and furnished below with large tufts of pale tomentum in the axils of the veins, and at maturity thin but firm in texture, dark green and very lustrous on the upper surface, paler and glabrous on the lower surface with the exception of a few scattered hairs along the slender midribs and remote primary veins extending obliquely toward the apex of the leaf, from 3 to 4.5 centimetres long and from 2 to 2.5 centimetres wide; stipules linear, glandular-serrate, caducous; petioles slender, grooved and slightly villose on the upper side at maturity and from 5 to 10 millimetres long; leaves on vigorous shoots usually broadly ovate, more or less deeply lobed above the middle and frequently from 4 to 5 centimetres long and from 3 to 4 centimetres wide. Flowers about 2 centimetres in diameter on elongated slender pedicels in three or four-flowered glabrous corymbs, with linear glandular red caducous bracts and bractlets; calyx-tube broadly obconic, glabrous, the lobes gradually narrowed, acuminate, entire, tipped with minute dark glands; stamens fifteen to twenty; anthers large, dark purple; styles five. Fruit subglobose to short-oblong, pendant on slender pedicels, 9 or 10 millimetres long, bright orange-red and marked by numerous pale dots; flesh thin, yellow, acidulous; nutlets five, about 6 millimetres long, rounded and obscurely ridged on the back.

A shrub, with slender wide-spreading stems from 4 to 8 metres in height, covered with close smooth bark varying from gray to brown, and slender branchlets at first orange-green, bright red-brown during their first season, gray tinged with red in their second year, marked by occasional small pale lenticels, and unarmed or occasionally furnished with slender straight spines rarely more than 3 centimetres in length. Flowers during the first week in March. Fruit ripens in June and early in July.

In low wet clay soil in a forest of Oaks and Palmettos bordering Haw Creek, Volusia County, Florida, *A. H. Curtiss* (No. 6677), July, 1900, March and April, 1901.

This is another species of the Æstivalis group, distinguished from *Cratægus maloides* by its smaller flowers in few-flowered corymbs, thinner leaves, and by its smaller and much less juicy fruits.

C. S. S.

EXPLANATION OF THE PLATE.

Plate VIII. Cratægus paludosa.

1. A flowering branch, natural size.
2. Vertical section of a flower, enlarged.
3. A calyx-lobe, enlarged.
4. A fruiting branch, natural size.
5. Vertical section of a fruit, enlarged.
6. Cross section of a fruit, enlarged.
7. A nutlet, side view, enlarged.
8. A nutlet, rear view, enlarged.

C. E. Faxon del.

CRATÆGUS LUCULENTA, Sarg.

EUPATORIUM LOESENERII, Robins.

Eupatorium Loesenerii, Robinson, *Proc. Am. Acad.* xxxv. 336 (1900).

Leaves opposite, petiolate, broadly ovate, entire, rounded or very obtuse at the apex, cordate or subcordate at the base, from 4 to 7 centimetres long, from 3 to 4.5 centimetres broad, thickish, subpalmately five to seven-nerved from a little above the base, pulverulent and grayish green above, fulvous tomentulose beneath, nerves and anastomosing veins prominent on the lower surface; petioles about 2 centimetres long, spreading, semiterete, deeply furrowed above, pulverulent-tomentulose; inflorescences dense many-headed rounded terminal corymbs; primary bracts petiolate, leaf-like, the ultimate ones much reduced, filiform; pedicels filiform, from 4 to 9 millimetres long; heads small, about 7 millimetres high, about twenty-flowered; involucre campanulate, scarcely half the length of the mature flowers; the scales oblong, obtuse, pulverulent, subequal, appressed; corollas 5 millimetres long, the short proper tube passing almost imperceptibly into a long slender tubular throat, the five-toothed limb relatively short, purplish. Achenes dark-colored, 2 millimetres long, columnar, tapering slightly toward the prominent callous base, the sharp angles slightly hispid.

A round-topped bush, from 1.8 to 2.8 metres high, with much-branched terete stems covered with firm gray longitudinally fissured bark, hard white wood surrounding a small white pith, opposite curved-ascending branches, buff-colored when young, and rather stout nearly terete obscurely striate branchlets clothed with very short almost pulverulent tomentum and leafy chiefly near the summit.

Mexico: calcareous hills, Oaxaca, along the Cuicatlan road, Valley of Oaxaca, at altitudes of from 2100 to 2400 metres, *E. W. Nelson* (No. 1549), October 3, 1894; Las Sedas, at an altitude of 1800 metres, *C. G. Pringle* (No. 6022, distributed as *E. sordidum*), October 30, 1894; San Juan del Estado, at an altitude of 1700 metres, *L. C. Smith* (No. 275), November 4, 1894; Cuauhtilla, *C. & E. Seler* (No. 1537), November 28, 1895.

It seems remarkable that this conspicuous species on the hills of Oaxaca was not observed by Galeotti or the other earlier explorers of the region, but was left for independent and almost simultaneous discovery by the collectors mentioned above. Concerning the country where it flourishes Mr. Pringle writes as follows: "At Las Sedas and westward the region takes the character of billowy hills, thinly covered with dwarfish Oaks and Pines. The rock formation is calcareous and the soil-covering scanty, consisting mainly of soft decomposing rock. Into this water has worn gulches and ravines innumerable, in whose shelter and on the steep bluffs above I gathered a long list of plants fully one third of which have proved to be new species. On these dry and chalky banks *Eupatorium Loesenerii* is at home in company with *Rhus mollis*, Humboldt, Bonpland & Kunth, *Dasylirion serratifolium*, Zuccarini, and such shrubby Salvias as thrive in none but these conditions. It forms a broad round-topped bush. With its numerous clusters of rosy flowers and leaves becoming purple-tinted with age and clad throughout in russet vesture this shrub is a pretty object. The species must be of frequent occurrence on similar hills about the Valley of Oaxaca and away to the westward through the elevated limestone region known as the Misteca Alta."

The species was dedicated to Dr. Theodor Loesener of the Royal Botanical Museum at Berlin.

B. L. Robinson.

Gray Herbarium.

EXPLANATION OF THE PLATE.

Plate IX. Eupatorium Loeseneri.

1. A flowering branch, natural size.
2. A head of flowers, enlarged.
3. A flower, enlarged.
4. A flower, corolla laid open, enlarged.
5. An achene, enlarged.

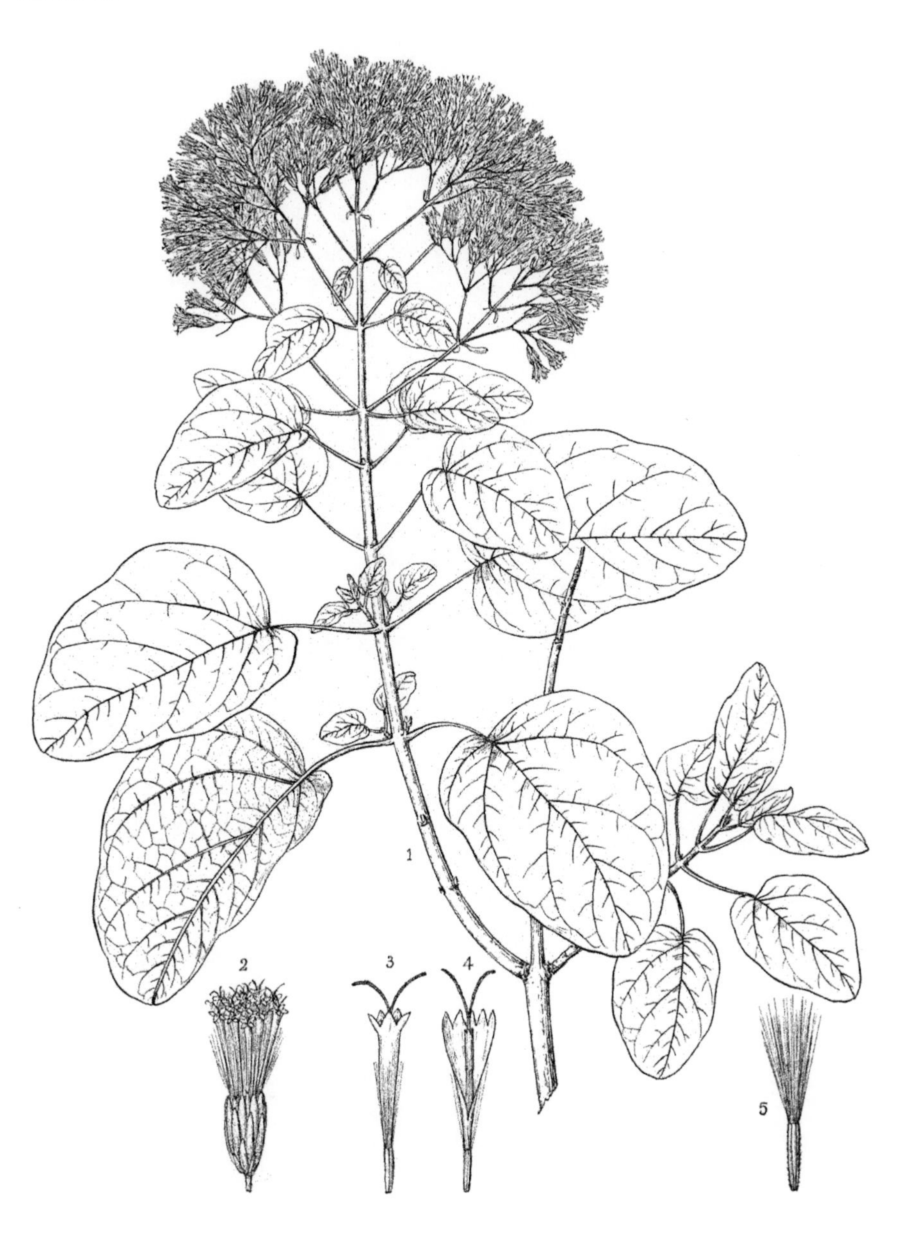

C. E. Faxon del.

EUPATORIUM LOESENERI, Robins.

SENECIO ROBINSONIANUS, Greenm.

Senecio Robinsonianus, Greenman, *Monogr. Senecio*, 26 (1901); *Engler Bot. Jahrb.* xxxii. 22.

Leaves long-petiolate, ovate-rotund, from 1.5 to 2 decimetres long, nearly or quite as broad, palmately seven to nine-nerved from or just above the more or less truncate or subcordate base, sinuately sublobate, with very shallow sinuses and mucronate-acute lobes, margined with numerous mucronulations, hirtellous-pubescent, densely and permanently lanate-tomentose beneath; petioles from 3.5 to 10 centimetres long, densely tomentose; inflorescence a terminal tomentose thyrsoidal panicle from 2 to 3 decimetres long and from 1 to 1.5 decimetres broad; heads about 8 millimetres high, radiate; involucre calyculate with small setaceous bracteoles more or less imbedded in a flocculent tomentum; bracts of the involucre commonly eight, lance-linear to lance-oblong, 4 millimetres long, obtusish, sparingly tomentulose, the inner scarious-margined; ray-flowers five; tubes slender, about equaling the pappus; rays short, 2 millimetres long, three or four-nerved; disk-flowers about eight. Achenes glabrous.

A shrubby perennial, from 2 to 3 metres high, with terete stems and densely matted tomentose branches.

Mexico: State of Oaxaca, between Nopala and Mixistepec, at altitudes of from 215 to 1230 metres above the sea-level, *E. W. Nelson* (No. 2439), March 5, 1895.

This species belongs to the Palmatinervii, a natural group of some thirty-four species, which in their geographical distribution extend from northern, or more particularly from north-central, Mexico to Costa Rica. The Palmatinervii are characterized by a more or less elongated axis of the inflorescence, and by having palmately nerved leaves. *Senecio Robinsonianus* is a typical representative of the group and is most nearly related to *Senecio roldana*, De Candolle, *Senecio Jaliscanus*, S. Watson, *Senecio albonervius*, Greenman, *Senecio Donnell-Smithii*, Coulter, and *Senecio lanicaulis*, Greenman. From these *Senecio Robinsonianus* is very distinct. The large ovate-rotund leaves, the elongated terminal paniculate inflorescence with the individual heads disposed in little glomerules, and the dense white tomentum of stem, branches, inflorescence, and under leaf-surface render it easily recognizable among all the numerous species of the genus. For the discovery of this species we are indebted to Mr. E. W. Nelson, an indefatigable collector, through whose rich and abundant collections of the past few years our knowledge of the flora of the more remote and less accessible parts of Mexico has been much increased.

Senecio Robinsonianus was named in honor of Professor B. L. Robinson; and specimens of the original, or type, are deposited in the National Herbarium at Washington, the Herbarium of the Royal Botanical Museum at Berlin, and in the Gray Herbarium at Cambridge.

J. M. Greenman.

Gray Herbarium.

EXPLANATION OF THE PLATE.

PLATE X. SENECIO ROBINSONIANUS.

1. A panicle of flowers, natural size.
2. Vertical section of a head, showing a disk-flower and a ray-flower, enlarged.
3. A disk-flower, enlarged.
4. A ray-flower, enlarged.
5. An involucre, enlarged.
6. A bract of an involucre, enlarged.
7. A leaf, natural size.

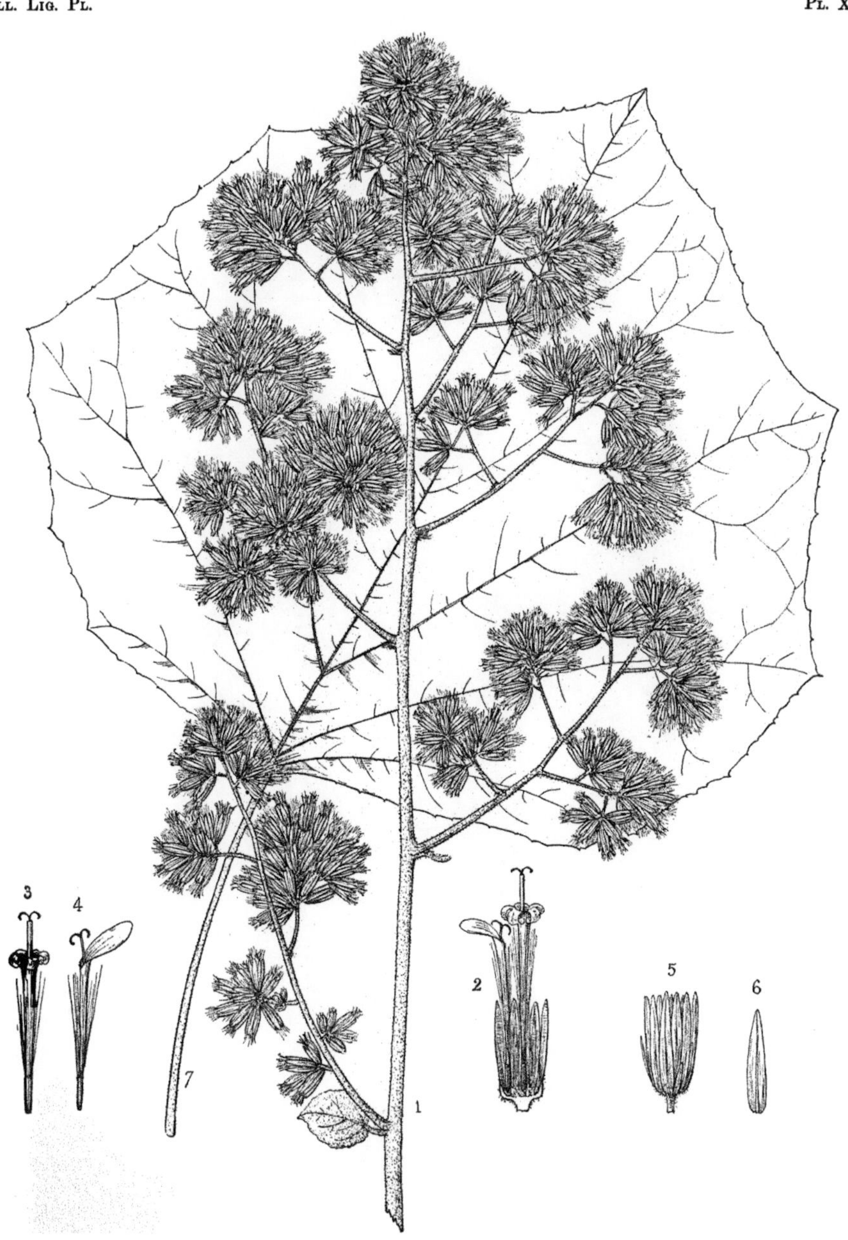

C. E. Faxon del.

SENECIO ROBINSONIANA, Greenm.

STYRAX RAMIREZII, Greenm.

Styrax Ramirezii, Greenman, *Proc. Am. Acad.* xxxiv. 568 (1899); xxxv. 309.

Leaves alternate, petiolate, oblong-lanceolate, from 1 to 1.5 decimetres long, from 3 to 6 centimetres broad, acuminate, acute or obtuse, entire, cuneate or rounded at the base, glabrous or essentially so above, paler and minutely stellate-lepidote beneath, the midrib and lateral veins prominent especially beneath, and more or less roughened by whitish or tawny stellate scales. Flowers in axillary secund racemes, 6 centimetres or less in length, including the peduncles, closely ferrugineous-stellate; calyx cupulate, from 5 to 6 millimetres high, about equaling the pedicels, shallowly sinuate, five-dentate, argenteous-lepidote, often with tawny stellate scales intermixed; corolla about 1.5 centimetres long, externally and along the margins of the inner surface of the lobes argenteous-lepidote; filaments and ovary stellate-pubescent. Fruit oblong, from 1 to 1.5 centimetres long, from 6 to 7 millimetres thick, pale green and closely stellate-lepidote over the entire surface.

A tree, from 9 to 12 metres high, the ultimate branches covered with a close ferrugineous-stellate tomentum.

Mexico: State of Morelos, in mountain cañons above Cuernavaca, at an altitude of 2000 metres, *C. G. Pringle* (No. 6848, fl. spec.), May 15, 1898; Sierra de Tepaxtlan, at an altitude of 2310 metres, *C. G. Pringle* (No. 8023, fr. spec.), February 7, 1899, and in the same locality (No. 9000, fl. spec.), May 6, 1900.

The statement in my original description, that the leaves are glabrous and smooth upon both surfaces, or slightly roughened on the prominent midrib and lateral nerves beneath, must be modified as above; and additional material shows that they are often somewhat broader than is indicated by the original specimen, and not unfrequently rounded as well as cuneate at the base.

The genus Styrax is represented in Mexico and Central America by ten known species, and of these *Styrax Ramirezii* seems most nearly related to *Styrax argenteum*, Presl, *Styrax contuminum*, Donnell-Smith, and the recently published *Styrax micranthum*, Perkins. In its geographical distribution *Styrax Ramirezii* is limited apparently to the plateau region of central Mexico, having been found thus far only in the State of Morelos, some sixty-four kilometres from the city of Mexico. As the species grows at rather high altitudes, it is not improbable that it could be successfully introduced into more northern latitudes. The light green foliage, the numerous axillary one-sided floral clusters, with somewhat drooping flowers, make it a most attractive tree.

Styrax Ramirezii was named in honor of Señor Dr. José Ramirez, Director of the Instituto Medico Nacional of the city of Mexico, one of the foremost naturalists of Mexico.

J. M. Greenman.

Gray Herbarium.

EXPLANATION OF THE PLATE.

Plate XI. Styrax Ramirezii.

1. A flowering branch, natural size.
2. A corolla laid open, enlarged.
3. A pistil, enlarged.
4. Portion of a cluster of fruit, natural size.
5. Vertical section of a fruit, enlarged.

C. E. Faxon del.

STYRAX RAMIREZII, Greenm.

FAXONANTHUS, Greenm.

Faxonanthus, *n. gen.*

Calyx 5(–6)-parted; segments narrow. Corolla broad, campanulate; lobes 5, rotund, somewhat unequal, the posterior lobe outermost in the bud, the anterior innermost. Stamens 4, in pairs, adnate to the base of the corolla, included; anther-cells divaricate, confluent, 1-celled. Ovary 2-celled, many-ovuled; style filiform; stigma subbilamellate. Capsule 2-valved; valves again dividing into two equal parts. Seeds small, pitted. Suffruticose plants with alternate or scattered leaves. Pedicels solitary in the axils. Flowers conspicuous.

Faxonanthus Pringlei, *n. sp.*

Leaves linear or spatulate-linear, 0.5 to 1.5 centimetres long, 1 to 3 millimetres broad, obtuse or acute, entire, glabrous and viscid, mostly present towards the free ends of the ultimately ascending minutely puberulent branches; the earlier leaves in falling leaving a raised elongated-elliptical subviscid portion of the originally petiolar base; pedicels short, 3 to 6 millimetres long; calyx about 8 millimetres high; corolla dark purple (or in the dried state somewhat blue), broadly campanulate, 2 to 2.5 centimetres high and equally broad, zygomorphic; corolla-tube anteriorly ventricose, pubescent and yellow-maculate at least in the dried state; calyx and the dehiscent capsule persistent long after the dissemination of the seeds.

A low shrub, 1 to 6 decimetres high and broad, with much-branched stems covered with a grayish bark.

Mexico: near Tehuacan, *C. G. Pringle* (No. 8594), 1901.

Faxonanthus is one of the most showy of the Mexican shrubs. Mr. Pringle, the collector, states that "it grows in clumps a foot or two in height and breadth on the white chalky hills near Tehuacan. With its fine dark green foliage and deep purple flowers the plant possesses a unique beauty."

The genus is named after the late Edwin Faxon, who, through a long period of botanical activity, did much to increase the knowledge of the flora of New England and to enrich many herbaria. I take pleasure in associating also with this plant the name of the well-known and sagacious collector, Cyrus G. Pringle.

J. M. Greenman.

Gray Herbarium.

EXPLANATION OF THE PLATE.

PLATE XII. FAXONANTHUS PRINGLEI.

1. A flowering branch, natural size.
2. A corolla laid open, enlarged.
3. A flower, the corolla removed, enlarged.
4. A pistil, enlarged.
5. A fruit, the calyx removed, enlarged.

C. E. Faxon del.

FAXONANTHUS PRINGLEI, Greenm.

EHRETIA VISCOSA, FERNALD.

EHRETIA VISCOSA, *n. sp.*

Leaves ovate, from 4.5 to 7.5 centimetres long, from 3 to 5 centimetres broad, subentire or crenate-serrate especially toward the short-acuminate tip, rounded or subcordate at the base, scabrous above, softly canescent beneath, the nerves and the petiole (from 6 to 12 millimetres long) densely glandular, panicles terminating the branchlets, from 4 to 5 centimetres high, nearly as broad, densely flowered; the rachis and pedicels viscid; calyx campanulate, puberulent, cleft half way to the base into broadly deltoid lobes; corolla white, 5 millimetres long, the exserted tube somewhat exceeding the recurved oblong lobes; stamens exserted, slightly exceeding the glabrous truncate style. Fruit oblong, 5 or 6 millimetres long, said to be white.

A large round-topped tree, about 12 metres high, with a trunk 1 metre or more in diameter, branches covered with smooth brown bark, and branchlets densely pubescent with soft spreading hairs, and viscid with intermixed brown glands.

Mexico: Morelos, near Cuernavaca, *C. G. Pringle* (No. 7777), May 29, 1899; *J. N. Rose & Walter Hough* (No. 4371), May 27–30, 1899; Guanajuato, ravine near the city of Guanajuato, *A. Dugès*, 1891.

This handsome broad-topped tree of the mountains of central Mexico resembles in its foliage and flowering panicles *Ehretia Mexicana*, S. Watson, and *Ehretia cordifolia*, Robinson, trees which were discovered by Mr. Pringle in western Mexico, and which are known only from the valleys of the Sierra Madre in the State of Jalisco. From both these species of western Mexico, *Ehretia viscosa* is easily distinguished by the glandular pubescence which is conspicuous upon the young portions of the tree, but is entirely wanting in *Ehretia Mexicana* and *Ehretia cordifolia*. *Ehretia viscosa* is apparently rare, and by a singular coincidence a solitary tree only is known to botanists at each of the stations from which our material has been collected. The tree from which the Cuernavaca specimens were collected is a very full and healthy specimen not far from the railroad. At Guanajuato the single tree known to Professor Dugès is a large specimen growing in a ravine near the city, where it was possibly introduced from farther south. This tree is known to the Mexicans as "Capulin blanco," because of the resemblance of its fruit to that of the Mexican Capulin, *Prunus serotina*, var. *salicifolia*, Koehne.

M. L. FERNALD.

Gray Herbarium.

EXPLANATION OF THE PLATE.

PLATE XIII. EHRETIA VISCOSA.

1. A flowering branch, natural size.
2. A flower, enlarged.
3. A corolla laid open, enlarged.
4. An anther, rear and front views, enlarged.
5. A flower, the corolla removed, enlarged.
6. Vertical section of a flower, the corolla removed, enlarged.
7. A fruiting branch, natural size.

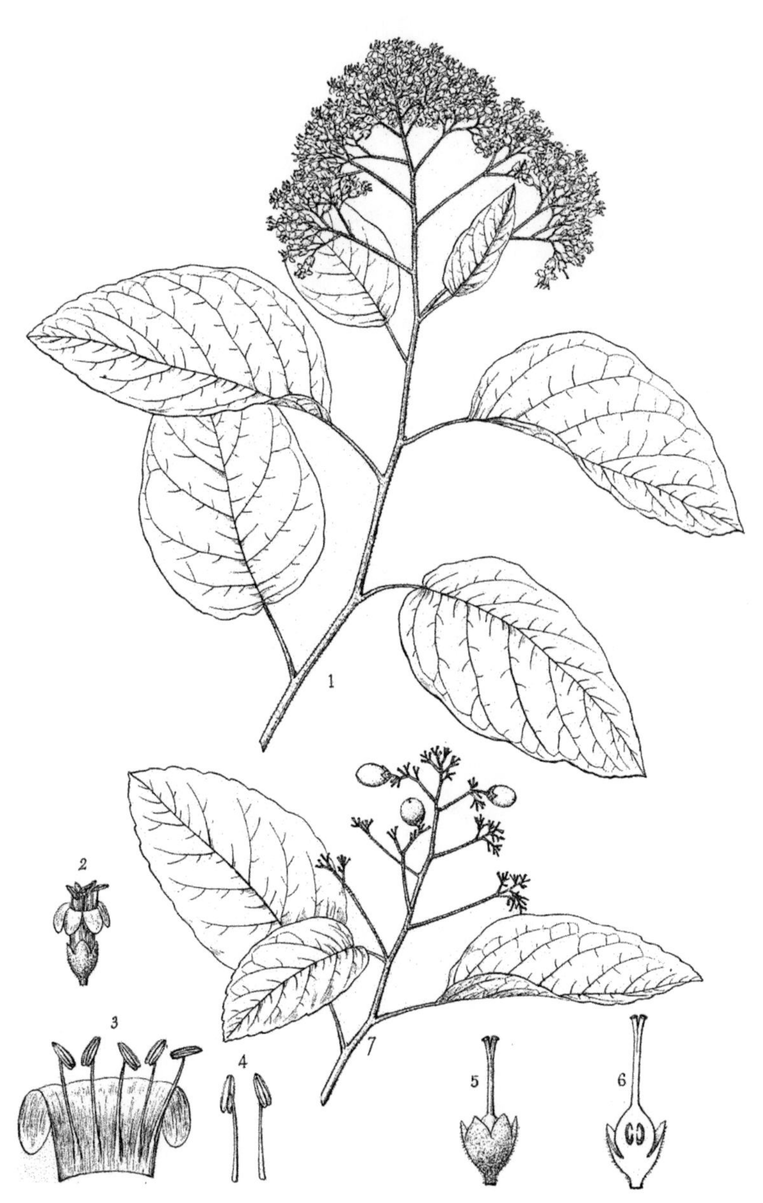

C. E. Faxon del.

EHRETIA VISCOSA, Fern.

BERBERIS SIEBOLDI, Miq.

Berberis Sieboldi, Miquel, *Ann. Mus. Lugd. Bat.* ii. 69 (1865–66); *Prol. Fl. Jap.* 1.—Regel, *Act. Hort. Petrop.* ii. 418. — Franchet & Savatier, *Enum. Pl. Jap.* i. 22. — Matsumura, *Fig. Descr. Pl. Koishikawa Bot. Gard.* ii. t. 5, A. — T. Ito, *Jour. Linn. Soc.* xxii. 427.—Koehne, *Deutsche Dendr.* 169.

Berberis vulgaris, Matsumura, *Fig. Descr. Pl. Koishikawa Bot. Gard.* ii. text to t. 5 (1883).

Leaves oblong-obovate to oblong or oblong-oblanceolate, from 3 to 7 centimetres long, acute or obtusish and mucronulate at the apex, gradually narrowed at the base into short petioles from 1 to 5 millimetres long, or sessile, densely and setosely ciliate and often revolute on the margins, dull green on the upper side, bright green and somewhat reticulate at maturity on the lower surface, deep vinous red in the autumn. Flowers appearing in May and June, pale yellow, about 8 millimetres in diameter, in almost umbel-like three to six-flowered peduncled racemes, nodding or spreading, shorter than the leaves; peduncles from 1 to 1.25 centimetres long; pedicels from 3 to 5 millimetres long, furnished at the base with an ovate-lanceolate bract, terminated by a bristle-like point, and from one third to one fifth as long as the pedicel; sepals six, the three outer orbicular-ovate, the inner longer and oval-ovate; petals deeply emarginate and usually incurved at the apex, about as long as the outer sepals; stamens about one third shorter than the petals, with filaments about 1.5 as long as the anthers; ovary ovoid with a sessile stigma. Fruits subglobose to ovoid, about 6 millimetres high, bright red and lustrous, rather dry, from two to five on a peduncle 1.5 to 2.5 centimetres long; seeds usually two, ovoid, 3 millimetres long, grayish brown, minutely punctulate, somewhat lustrous.

A low broad roundish shrub, from 30 to 75 centimetres high, with numerous upright or spreading stems clothed with light grayish brown longitudinally fissured bark, the branches of the previous year dark reddish brown, angular, compressed and two-edged toward the apex, and armed with three-parted slender brown spines sometimes 12 millimetres long.

Japan: in the mountainous regions of Hokkaido and Hondo (ex Ito); Hondo, on the Nagasendo near Metaki, *C. S. Sargent*, 1892.

Berberis Sieboldi seems most nearly allied to *Berberis umbellata*, Wallich, and to *Berberis serrata*, Koehne, but is easily distinguished by its setosely ciliate leaves. According to T. Ito *Berberis Sieboldi* is the Hirohano Hebinoborazu of Keiske Ito (*Somoku Dsusetsu Arb.* ii. t. 5, B). It is one of the most distinct of the cultivated Barberries on account of its foliage and habit. The purplish color of the unfolding leaves marked with green veins is specially noticeable, and the foliage is also very handsome in autumn, when it turns to a bright vinous red. Like that of *Berberis Thunbergi*, De Candolle, the rather dry fruit retains its bright color unchanged until the following spring. *Berberis Sieboldi* was introduced into the Arnold Arboretum by seeds collected in 1892 by Professor Sargent in Japan. It has proved perfectly hardy in the vicinity of Boston.

Alfred Rehder.

Arnold Arboretum.

EXPLANATION OF THE PLATE.

PLATE XIV. BERBERIS SIEBOLDI.

1. A flowering branch, natural size.
2. Vertical section of a flower, enlarged.
3. A bractlet from the base of the flower, enlarged.
4. An outer sepal, enlarged.
5. An inner sepal, enlarged.
6. A petal, enlarged.
7. A stamen, enlarged.
8. A pistil, enlarged.
9. A fruiting branch, natural size.
10. Vertical section of a fruit, enlarged.
11. A seed, enlarged.

C. E. Faxon del.

BERBERIS SIEBOLDI, Miq.

ILEX SERRATA, Thunb.

Ilex serrata, Thunberg, *Fl. Jap.* 78 (1784). — De Candolle, *Prodr.* ii. 16. — Franchet & Savatier, *Enum. Pl. Jap.* i. 78. — Rehder, *Cycl. Am. Hort.* ii. 798; *Mitt. Deutsche Dendr. Gesell.* x. 111. — Loesener, *Nov. Act. Nat. Car.* lxxviii. 467 (*Monogr. Aquifol.*).

Ilex Sieboldi, Miquel, *Versl. Med. Kon. Akad. Weten.* ser. 2, ii. 84 (1866); *Ann. Mus. Lugd. Bat.* iii. 104; *Prol. Fl. Jap.* 268. — Franchet & Savatier, *Enum. Pl. Jap.* i. 77. — Maximowicz, *Mém. Acad. Sci. St. Pétersbourg*, sér. 7, xxix. No. 3, 48. — Nakagawa, *Tokyo Bot. Mag.* xiii. 108. — Shirasawa, *Icon. Ess. For. Jap.* i. 101, t. 61.

Ilex serrata var. Sieboldi, Rehder, *Cycl. Am. Hort.* ii. 798 (1900); *Mitt. Deutsche Dendr. Gesell.* x. 111. — Loesener, *Nov. Act. Car.* lxxviii. 468 (*Monogr. Aquifol.*).

Leaves deciduous, elliptic or ovate to ovate or obovate-oblong, rarely lanceolate, 2.5 to 8 centimetres long, rarely smaller, acuminate, narrowed at the base into glabrous or puberulous petioles from 4 to 10 millimetres long, finely and densely serrate, with spreading acuminate teeth, on the upper surface dull green and glabrous, or pubescent along the midrib, with short hairs, lighter green and finely pubescent or glabrous on the lower surface, with from six to nine prominent secondary veins impressed above. Flowers appearing in June and July in the axils of the leaves and of the basal bracts of the branches, diœcious; staminate flowers generally 4-merous with a conic pistillode, in simple to thrice dichotomous, from three to seven or rarely to fifteen-flowered, short-pedunculate or sessile cymes; pedicels slender, from 0.5 to 1.5 millimetres long, puberulous or glabrous; pistillate flowers generally 5-merous, with staminodes nearly like stamens, disposed in simple or twice dichotomous, from one to three or rarely to seven-flowered short-pedunculate or sessile cymes; pedicels from 0.5 to 2.5 millimetres long, puberulous or glabrous: flowers of both sexes hardly differing in size, from 4 to 5 millimetres in diameter; calyx minutely hirsute, broadly cup-shaped, with four ovate triangular ciliolate lobes, rounded or subacute, as long as the tube in the pistillate, longer in the staminate flowers; petals about thrice as long as the calyx-teeth, connate at the base, ovate in the pistillate, oval in the staminate flower, whitish, or tinged with pink particularly on the outer surface; stamens somewhat shorter than the petals; anthers ovoid, almost twice as long as the free part of the filaments, in the pistillate flowers cordate-ovate, longer than the free part, but shorter than the whole of the filament; ovary conic-ovoid with a sessile four, rarely five or six-lobed stigma, four or rarely five to six-celled; ovule solitary. Fruit a globose drupe ripening in October, bright red, from 4 to 6 millimetres in diameter; stones usually four, from 2.5 to 3.2 millimetres long, ovoid-oblong, obtusely trigonous, yellowish white, smooth; seed light brown, minutely undulate-rugulose.

A shrub or small tree, often 4 or 5 metres high, or occasionally taller, with a trunk scarcely attaining 20 or exceptionally 30 centimetres in diameter and covered with smooth bark, dark gray or greenish gray spreading or upright branches, and slender somewhat zigzag branchlets puberulous when young or glabrous.

Japan: in the lower mountainous regions of Hondo mostly near the borders of streams and in low ground; and in Kiu-shiu.

Besides the type which has been distinguished as *Ilex serrata*, var. *Sieboldi*, the following well-marked varieties have been described: —

Ilex serrata, var. *argutidens*, Rehder, *Cycl. Am. Hort.* ii. 798 (1900); *Mitt. Deutsche Dendr. Gesell.* x. 111. — Loesener, *Nov. Act. Nat. Car.* lxxviii. 468 (*Monogr. Aquifol.*).

Ilex argutidens, Miquel, *Versl. Med. Kon. Akad. Weten.* ser. 2, ii. 84 (1866); *Ann. Mus. Lugd. Bat.* iii. 104; *Prol. Fl. Jap.* 268. — Franchet & Savatier, *Enum. Pl. Jap.* i. 76.

Ilex serrata, Maximowicz, *Mém. Acad. Sci. St. Pétersbourg*, sér. 7, xxix. No. 3, 48 (1881). — Yatabe, *Tokyo Bot. Mag.* vi. 158.

Branchlets and leaves glabrous, the latter usually smaller; flowers oftener 4-merous, pinkish. In Hondo; not reported from Kiu-shiu.

Ilex serrata, var. *subtilis*, Loesener, *Nov. Act. Nat. Car.* (*Monogr. Aquifol.*) lxxviii. 469 (1901).

Ilex subtilis, Miquel, *Versl. Med. Kon. Akad. Weten.* ser. 2, ii. 84 (1866); *Ann. Mus. Ludg. Bat.* iii. 107; *Prol. Fl. Jap.* 271. — Franchet & Savatier, *Enum. Pl. Jap.* i. 78.

Ilex Sieboldi, var. *subtilis*, Yatabe, *Tokyo Bot. Mag.* vi. 158 (1892).

A dwarf form with lanceolate leaves about 1.5 centimetres long and 0.4 centimetres broad or even smaller, pubescent beneath; flowers smaller, usually 4-merous. Cultivated in Japan.

Ilex serrata is closely allied to *Ilex verticillata*, Gray, which differs in its larger and more coarsely serrate leaves, in the usually 6 to 9-merous flowers and in the size of the fruit, which is almost twice as large. Like *Ilex verticillata*,[1] the Japanese species has produced forms with yellow and also with whitish berries.[2] On account of its attractive fruit, which remains a long time on the branches after the leaves have fallen, *Ilex serrata* is a favorite plant with the Japanese, who often plant it for ornament. In autumn the leafless branches loaded with the bright red berries are brought into the cities and sold in the streets for the decoration of dwelling-houses.[3]

Ilex serrata has been cultivated in the Arnold Arboretum for many years and has proved perfectly hardy. It seems to be still little known in European gardens. It was first introduced into those of the United States about 1866 by Thomas Hogg, while the var. *argutidens* was raised from seeds collected by Professor Sargent in Japan in 1892. The second variety is probably not yet in cultivation outside of Japan. As an ornamental plant *Ilex serrata* is chiefly valued for the effect which the attractive berries produce after the leaves have fallen; and though less showy, it is more graceful than the better known *Ilex verticillata*.

Alfred Rehder.

Arnold Arboretum.

1 Robinson, *Rhodora*, ii. 106. This form has been in cultivation in the Arnold Arboretum since 1885.
2 Bean, *Gard. Chron.* ser. 3, xxviii. 322.
3 Sargent, *Garden & Forest*, vi. 122; *For. Fl. Jap.* 25.

EXPLANATION OF THE PLATE.

Plate XV. Ilex serrata.

1. A branch with staminate flowers, natural size.
2. A tetramerous staminate flower, enlarged.
3. A pentamerous staminate flower, enlarged.
4. Vertical section of a staminate flower, enlarged.
5. A branch with pistillate flowers, natural size.
6. A pistillate flower, enlarged.
7. Vertical section of a pistillate flower, enlarged.
8. A fruiting branch, natural size.
9. A fruit divided transversely, enlarged.
10. A seed, enlarged.

C. E. Faxon del.

ILEX SERRATA, Thunb.

ACER CAPILLIPES, MAXIM.

ACER CAPILLIPES, Maximowicz, *Bull. Acad. Sci. St. Pétersbourg*, xii. 225 (1867); xxvi. 441; *Mél. Biol.* vi. 367; x. 597. — Miquel, *Arch. Néerland.* ii. 477. — Franchet & Savatier, *Enum. Pl. Jap.* i. 88. — Pax, *Bot. Jahrb.* vii. 246; *Engler Pflanzenreich*, 8 Heft (iv. 163) 67. —Rehder, *Mitt. Deutsche Dendr. Gesell.* x. 115.

ACER PENNSYLVANICUM, var. CAPILLIPES, Wesmael, *Bull. Soc. Bot. Belg.* xxix. 62 (1890).

Leaves long-petioled, membranaceous, glabrous, orbicular-ovate, from 8 to 12 centimetres long, from 6 to 8 centimetres broad, rounded or subcordate at the base, three or rarely five-lobed, sharply and doubly serrate, the middle lobe much larger than the others, triangular-ovate, acuminate, the lateral lobes very short, acuminate, spreading outward, the upper surface dark green, opaque, the lower surface light green usually with reddish veins; petioles slender, from 3 to 5 centimetres long, glabrous, usually red. Flowers appearing with the leaves; racemes short-peduncled, slender, drooping, from 7 to 10 centimetres long, many-flowered, glabrous; pedicels filiform, from 8 to 15 millimetres long; flowers greenish white, 8 or 9 millimetres in diameter; sepals spatulate, 3 millimetres long; petals obovate, 4 millimetres long; stamens perigynous, shorter than the petals; filaments glabrous; disk rather small. Fruits glabrous, about 3 millimetres long, light reddish or yellowish brown, numerous in pendulous racemes; wings straight, rounded at the apex, 0.5 centimetres broad, spreading at a very obtuse angle or almost horizontal.

A tree, about 10 metres in height, with a trunk sometimes 30 centimetres in diameter, erect greenish brown branches conspicuously marked by pale longitudinal stripes, glabrous branchlets usually tinged more or less with purple and sometimes slightly covered toward the apex with a glaucous bloom, and stalked winter-buds with two pairs of valvate scales, those of the outer pair glabrous and inclosing the densely silky-pubescent scales of the inner pair.

Japan: Hondo, Province of Shenano, *Tschonoski*, 1864 (in flower), banks of the Kisogawa near Agamatsu on the Nagasendo, *C. S. Sargent*, October 5, 1892 (fruit).

Acer capillipes is distinguished by its glabrous leaves from all other three-lobed species of the section Macracantha of Pax, except from *Acer tegmentosum*, Maximowicz, and *Acer pectinatum*, Wallich, which also have glabrous foliage, but rather short-pediceled flowers and fruits. *Acer pectinatum* may be distinguished by the setosely pointed teeth and the longer lateral lobes pointing forward of the leaves, and *Acer tegmentosum* by the lateral lobes of the leaves pointing forward and the green color of the unfolding leaves, while in *Acer capillipes*, at least in the cultivated plants, the young unfolding foliage is bright red.

Acer capillipes, which appears to have been collected only twice, and is probably extremely rare, is a handsome tree similar in habit to *Acer Pennsylvanicum*, Linnæus, and *Acer rufinerve*, Siebold & Zuccarini, but easily distinguished from them in spring by the bright red color of the unfolding leaves, which is usually retained through the season on the branchlets, petioles and the veins, and adds much to the beauty of the plant. Late in autumn the leaves assume a dark purple color. *Acer capillipes* has proved hardy at the Arnold Arboretum, where it was introduced in 1892 by Professor Sargent, who collected seeds in Japan.

ALFRED REHDER.

Arnold Arboretum.

EXPLANATION OF THE PLATE.

PLATE XVI. ACER CAPILLIPES.

1. A flowering branch, natural size (from specimens collected by Tschonoski and now in the Gray Herbarium).
2. A staminate flower, enlarged.
3. A sepal, enlarged.
4. A petal, enlarged.
5. A stamen, enlarged.
6. A fruiting branch, natural size (collected by Sargent).
7. Vertical section of a nut, enlarged.
8. An embryo, enlarged.

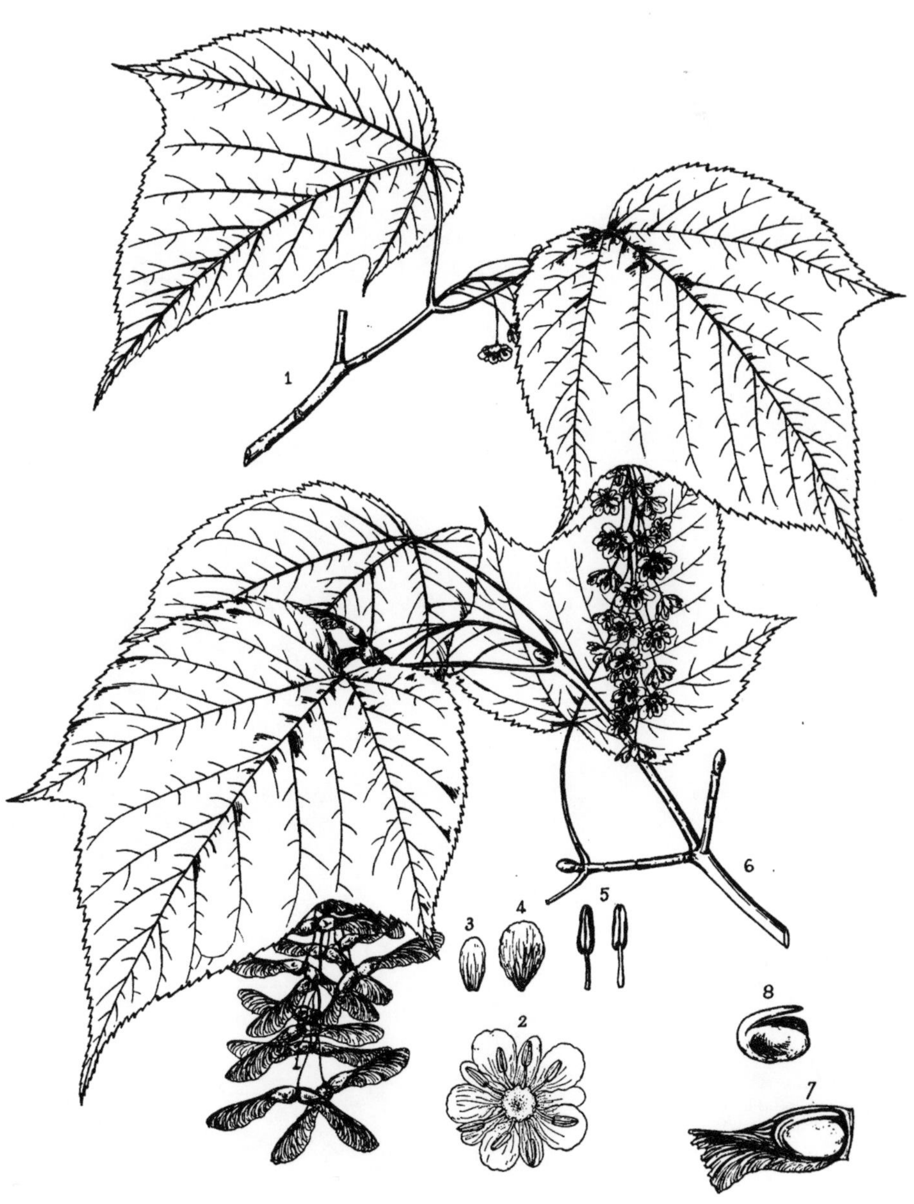

C. E. Faxon del.

ACER CAPILLIPES, Maxim.

ACER TSCHONOSKII, MAXIM.

ACER TSCHONOSKII, Maximowicz, *Bull. Acad. Sci. St. Pétersbourg,* xxxi. 24 (1886); *Mél. Biol.* xii. 432. — Pax, *Bot. Jahrb.* xi. 80; *Engler Pflanzenreich,* 8 Heft (iv. 163) 70.

ACER CAPILLIPES, Sargent, *Garden and Forest,* vi. 153 (not Maximowicz) (1893); *Forest Fl. Jap.* 30.

Leaves slender-petioled, membranaceous, orbicular-ovate, 4 to 10 centimetres long, cordate at the base, deeply five or rarely seven-lobed, the lobes reaching about half way to the middle, triangular-ovate, long-acuminate, incisely and sharply double-serrate, with incurved teeth acuminate with setose tips, bright green and glabrous on the upper surface, lighter green beneath and rufously villose along the primary veins, almost glabrous at maturity except in the axils of the veins near the base of the leaf; petioles slender, from 2 to 5 centimetres long, occasionally longer, rufously villose near the base of the blade. Flowers appearing with the leaves; racemes from six to ten-flowered, from 1 to 1.5 centimetres long; peduncle from 4 to 5.5 centimetres long, glabrous; pedicels filiform, the lower 1 centimetre long, the upper shorter; flowers from 8 to 10 millimetres in diameter, greenish white; sepals oblong, about one third shorter than the oblong-obovate petals; staminate flowers with stamens slightly shorter than the petals, the glabrous filaments many times longer than the oval anthers; disk intrastaminal with a rudimentary pistil in the middle; pistillate flowers with rudimentary short stamens; pistil glabrous, the style much shorter than the spreading slender stigmas. Fruits light yellowish brown, in racemes of three to six; pedicels filiform, from 1 to 1.5 centimetres long; wings spreading at an obtuse or almost right angle, each samara from 2 to 2.3 centimetres long, from 8 to 10 millimetres broad, the wing incurved and almost truncate at the apex.

A shrub or small tree, about 5 metres high, with gray or grayish brown smooth-barked branches, glabrous greenish branchlets usually tinged with purple, and small stalked winter-buds covered by two pairs of valvate scales, the outer glabrous and entirely inclosing the inner, which are silky-pubescent.

Japan: in the mountain woods of Hondo up to altitudes of 800 or 900 metres, *Tanaka, Tschonoski, Rein, Faurie* (ex Pax); Mt. Hakkoda, near Aomori, up to 1400 metres, *C. S. Sargent,* October 3, 1892; Uzen, Herb. Imp. Univ. Tokyo; Hokkaido, Raiden Mount, Province of Shiribeshi, *K. Sugiyama,* July, 1888.

Acer Tschonoskii is closely related to *Acer micranthum,* Siebold & Zuccarini, which differs from it chiefly in its longer many-flowered racemes, and in its shorter and stouter pedicels, while the sepals are only about half as long as the obovate petals, and the under sides of the leaves are glabrous with the exception of the tufts in the axils of the veins. *Acer Tschonoskii* is a shrubby tree with small graceful foliage somewhat similar in appearance to *Acer palmatum,* Thunberg. It is proved hardy at the Arnold Arboretum, where it was introduced in 1892 by Professor Sargent, who collected seeds in Japan, and where it flowered this year for the first time. Seeds of this species were first sent from the Arboretum to some European gardens under the name of *Acer capillipes.*

ALFRED REHDER.

Arnold Arboretum.

EXPLANATION OF THE PLATE.

Plate XVII. Acer Tschonoskii.

1. A flowering branch, natural size.
2. A staminate flower, enlarged.
3. Vertical section of a pistillate flower, enlarged.
4. A fruiting branch, natural size.
5. Vertical section of a nut.

C. E. Faxon del.

ACER TSCHONOSKII, Maxim.

MALUS HALLIANA, Koehne.

Malus Halliana, Koehne, *Gatt. Pomaceen*, 27 (1890); *Deutsche Dendr.* 261.

Pyrus Halliana, Voss, *Vilmorin Blumeng.* ed. 3, i. 277 (1896).

Pyrus Parkmani, *Hort.*

Leaves convolute in the bud, petioled, ovate-oblong to elliptic-oblong, from 4 to 7 centimetres long, acute or somewhat acuminate, narrowed or rounded at the base, serrulate, with appressed obtusish teeth, chartaceous at maturity, dark green, with impressed veins, lustrous and glabrous above with the exception of the hairy and glandular midrib, lighter green and glabrous below usually with a purplish midrib; petioles from 1 to 1.2 centimetres long, pubescent when young, glabrous at maturity. Flowers andro-polygamous, appearing with the leaves in from two to six-flowered umbels, rose-colored, from 3.5 to 4 centimetres in diameter; pedicels glabrous, from 2.5 to 4 centimetres long, like the calyx dark purple; sepals ovate, glabrous on the outer surface, villose within, about one half as long as the calyx-tube; petals ovate, rounded, sparingly villose at the unguiculate base and often with scattered hairs on the inner surface toward the apex; stamens about twenty, of unequal length, the longest about two thirds the length of the petals; filaments glabrous; styles four or five, connate almost to the middle and densely villose below the middle, glabrous at the base. Fruit subglobose, about 6 millimetres in diameter, dark purple, four or five-celled, abruptly contracted at the base into the slightly thickened straight spreading pedicel and crowned by the small scar of the deciduous calyx deciduous before the fruit ripens; seeds broadly ovate, slightly compressed, 4 millimetres long.

A small tree, from 3 to 5 metres high, often shrubby, with spreading branches forming a broad open head, a short stem covered with smooth gray bark, and slender usually dark purple branchlets sparingly hairy at first and soon glabrous.

Japan: probably only in gardens; the only Japanese specimen I have seen is from the Tokyo Botanical Garden, collected by *K. Miyabe* and called *Pyrus spectabilis*. According to Koehne *Malus Halliana* was also collected by *Hilgendorf* and by *Maximowicz*, and by the latter distributed as *Cratægus alnifolia*, Siebold & Zuccarini, but a fruiting specimen from the collection of Maximowicz under this name in the Gray Herbarium certainly belongs to *Malus Toringo*, Siebold. Like many other plants cultivated in Japanese gardens, it may have been introduced from China, although it does not seem to have yet been collected there.

Malus Halliana belongs to the subsection Gymnomeles and is allied to *Malus baccata*, Desfontaines, and *Malus floribunda*, Siebold; from the first it is distinguished by the leathery leaves, the color of the flowers, the much shorter sepals, the purple calyx and pedicels, and the four or five-celled very late ripening fruit; from *Malus floribunda* it is distinguished by the convolute vernation of the glabrous leaves, the color of the larger flowers, the shorter sepals, and the glabrous purple pedicels and calyx. In foliage and flowers it much resembles *Malus spectabilis*, Desfontaines, which, however, differs by its pubescence and the much larger fruit crowned by the persistent calyx.

From all other species of *Malus*, and in fact from all Sorbæ, except *Chænomeles Japonica*, Lindley (*Pyrus Japonica*, Thunberg), *Malus Halliana* differs in its polygamous flowers, which seem to have hitherto escaped notice. There is at least one staminate flower in each umbel, and this is always terminal; sometimes there are two or three, but the number of staminate flowers rarely exceeds that of the perfect ones. In the staminate flowers there is no trace of reduced pistils, which are present, though often very minute, in the sterile flowers of *Chænomeles Japonica*.

Malus Halliana is one of the most beautiful of the Asiatic Crab-apples. The delicate color of the large flowers gracefully nodding on slender dark-colored pedicels give it a very distinct and charming appearance; it is also handsome in summer with its dark glossy foliage, which remains on the branches until late in autumn, when it assumes a dark purple color. *Malus Halliana* was first introduced about 1863 into American gardens by Dr. G. R. Hall, an American physician, who resided for many years in China and Japan. The first mention of this beautiful Crab-apple seems to be in a note in the *Rural New Yorker*[1] of 1882 under the name of *Malus Halliana;* eight years afterwards a photograph of a flowering branch was published in *Garden and Forest*,[2] but without technical description. A photograph of a flowering tree showing the habit together with that of a flowering branch may be found in Möller's *Deutsche Gärtner-Zeitung* of 1899.[3] In the vicinity of Boston *Malus Halliana* has proved hardy, but Massachusetts seems to be the northern limit of its hardiness in the eastern states.

ALFRED REHDER.

Arnold Arboretum.

[1] See *The Garden*, xxii. 162 (1882).
[2] Sargent, *Gard. & For.* i. 152, f. 30 (1888).
[3] Rehder, *Möller Deutsche Gärtner-Zeit.* xiv. 457, f. (1899).

EXPLANATION OF THE PLATE.

PLATE XVIII. MALUS HALLIANA.

1. A flowering branch, natural size.
2. A flowering branch, natural size (from a Japanese specimen).
3. Vertical section of a staminate flower, enlarged.
4. Vertical section of a perfect flower, enlarged.
5. A sepal, enlarged.
6. A petal, enlarged.
7. A fruiting branch, natural size.
8. Vertical section of a fruit, enlarged.
9. Cross section of a fruit, enlarged.
10. A seed, enlarged.

C. E. Faxon del.

MALUS HALLIANA, Koehne

VIBURNUM WRIGHTII, MIQ.

VIBURNUM WRIGHTII, Miquel, *Ann. Mus. Lugd. Bat.* ii. 267 (1865–66); *Prol. Fl. Jap.* 155. — Maximowicz, *Bull. Acad. Sci. St. Pétersbourg*, xxvi. 490; *Mél. Biol.* x. 667. — Nakagawa, *Tokyo Bot. Mag.* ii. 76. — Rehder, *Cycl. Am. Hort.* iv. 1926.

VIBURNUM EROSUM, Gray, *Perry Exped.* ii. 313 (not Thunberg) (1856).

Leaves exstipulate, petioled, orbicular to broadly ovate or obovate, from 6 to 14 centimetres long, abruptly acuminate, rounded or broadly cuneate at the base, coarsely and sinuately dentate, with from twelve to twenty, or rarely twenty-five apiculate teeth on each side, and with from six to eight pairs of almost parallel lateral veins; bright green on the upper surface and lighter green below, with short hairs on the upper side of the veins and with scattered long hairs on the lower side, otherwise glabrous or furnished with scattered simple or fasciculate hairs on both sides when young; petioles glabrous or covered with long scattered hairs, from 8 to 18 millimetres long. Flowers appearing in June in upright pedunculate cymes; peduncles from 1.5 to 3 millimetres long, hirsute; cymes five-rayed, from 5 to 10 centimetres broad, finely pubescent and glandular; flowers sessile or short-pediceled, with small caducous bracts at the base, arranged in raylets of the third and fourth order; calyx-teeth glabrous, orbicular-ovate, short; corolla rotate, from 5 to 6 millimetres in diameter, glabrous, with orbicular-ovate lobes; stamens exceeding the corolla; style short and thick, with a two-lobed stigma. Drupe subglobose, scarlet, about 1 centimetre high, juicy; stone broadly ovate, pointed, much flattened, 6 millimetres broad, yellowish white, rugulose and furnished with five fine longitudinal lines. Seed brown, the testa minutely punctulate.

An erect shrub, attaining a height of 3 metres, with upright spreading grayish brown branches, slender branchlets which are glabrous or furnished with scattered spreading hairs, and rather large winter-buds covered by two pairs of imbricate scales which are yellowish pubescent toward the apex, or those of the outer pair glabrous.

Japan: Hokkaido, near Hakodate, *C. Wright*, 1853–56; *Albrecht*, 1860; *C. S. Sargent*, near Kakumi Hot Springs, September 29, and Sapporo, September 17, 1892; Sapporo, *E. Tokubuchi*, September 8, 1894; Hondo, Mountains of Niko, *Tschonoski*, 1864; on Fuji-san, *K. Watanabe*, July 29, 1891; Chokaizan, *J. H. Veitch*, September 15, 1892; near Fukishima, on the Nagasendo, *C. S. Sargent*, October 25, 1892; Kiu-shiu (ex Nakagawa). China: Szechuen, *Henry* (No. 6262).

Viburnum Wrightii is closely allied to *Viburnum dilatatum*, Thunberg, and to *Viburnum phlebotrichum*, Siebold & Zuccarini. From *Viburnum dilatatum* it is chiefly distinguished by its almost glabrous leaves, glabrous flowers, and the larger and more juicy fruits. *Viburnum phlebotrichum*, which is similar in pubescence and in fruit, has much smaller and narrower leaves, with stamens shorter than the corolla and smaller slender-stemmed glabrous cymes with purplish peduncles and calyx. Hemsley [1] united *Viburnum Wrightii* with *Viburnum phlebotrichum*, saying that Szechuen specimens connect those two species. According to specimens which I have before me, both species seem to grow in Szechuen, but are readily distinguished, although the Chinese specimens of *Viburnum phlebotrichum* have longer stamens than those from Japan and resemble in this respect *Viburnum Wrightii*. From the other localities mentioned by Hemsley I have seen no specimens. *Viburnum Willeanum*, Graebner,[2] appears to differ little from *Viburnum Wrightii* as here described. Dr. Henry's specimen from Szechuen quoted above agrees well with Graebner's description in regard to the large inflorescence, which is 10 centimetres in diameter, and to the smaller flowers,

[1] Forbes & Hemsley, *Jour. Linn. Soc.* xxiii. 354.

[2] Graebner, *Bot. Jahrb.* xxix. 589 (*Flora von Central China*).

but the leaves are not obovate, and are entire only in the lower third, while those of *Viburnum Willeanum* are described as rhombic-obovate, sinuately dentate only above the middle and attaining a length of but 8 centimetres.

Viburnum Wrightii is a handsome shrub, especially ornamental in autumn with its large clusters of glossy scarlet fruits resembling those of *Viburnum Opulus*, Linnæus, although somewhat smaller. It is also attractive when in flower and when the foliage turns to a dark red in autumn. It was introduced from Japan in 1892 by Professor Sargent into the Arnold Arboretum, where it has proved hardy in a sheltered position or with some protection during the winter.

ALFRED REHDER.

Arnold Arboretum.

EXPLANATION OF THE PLATE.

PLATE XIX. VIBURNUM WRIGHTII.

1. A flowering branch, natural size.
2. A flower with the corolla opened, enlarged.
3. A fruiting branch, natural size.
4. A stone, enlarged.
5. Cross section of a stone, enlarged.

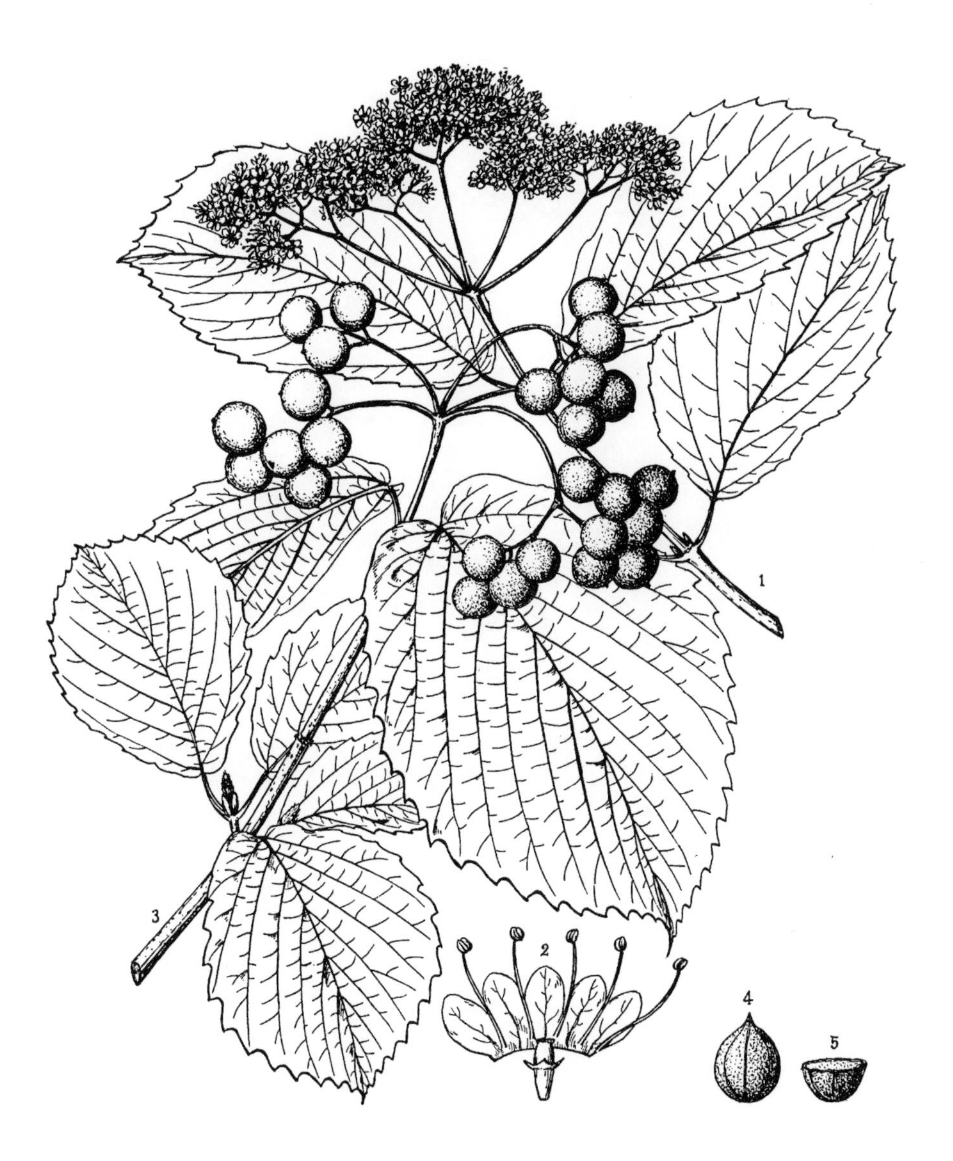

C. E. Faxon del.

VIBURNUM WRIGHTII, Miq.

LONICERA SACCATA, Rehd.

Lonicera, *n. sp.*

Lonicera No. 32 (Xylostum) *sp. nov.?* Hemsley, *Jour. Linn. Soc.* xxiii. 368 (1888).

Lonicera microphylla, Graebner, *Bot. Jahrb.* xxix. 595 (*Flora von Central China*) (1901).

Leaves elliptic-oblong or rhombic-oblong to oblong, from 1.5 to 5 centimetres long, from 6 to 15 millimetres broad, obtuse, narrowed at the base into an almost glabrous or sparingly hairy petiole from 1 to 2 millimetres in length, blade dark green and sparingly pubescent on the upper surface, sometimes only along the midrib, glaucescent on the lower surface, and pubescent when young, becoming glabrous at maturity, except on the rather prominent yellowish veins. Flowers in pairs on glabrous slender nodding peduncles from 1 to 2.5 centimetres long, in the axils of the lower leaves; bracts narrowly oblong, acute, sparingly ciliate, exceeding the calyx; bractlets none; ovaries wholly connate, each ovary three-celled, cells three-ovuled; calyx small, cup-shaped, indistinctly five-toothed, sparingly ciliate or wholly glabrous; corolla tubular, 10 millimetres long, slightly and gradually widened from the base, apparently white suffused with pink, glabrous on the outer surface, hairy within, limb short, about 2 millimetres long, with 5 ovate upright lobes; stamens inserted in the upper third of the tube, about as long as the limb; filaments glabrous, scarcely longer than the linear-oblong anthers; style exserted, villose below the middle with long scattered hairs. Fruit a subglobose berry consisting of the two connate ovaries, scarlet, from 6 to 8 millimetres in diameter, many- (to eighteen-) seeded; seeds oval, 4.5 millimetres long, compressed, smooth, light brown.

An upright shrub from 0.5 to 1.5 metres high, with slender branches covered with light grayish brown shredding bark, glabrous young branchlets more or less purple, and small winter-buds covered by several pairs of lanceolate minute scales, the innermost foliaceous.

China: Sze-chuen, *Henry* (No. 5680, 5680 A); Mount Omei, *Faber* (No. 66); Hupeh, *Henry* (No. 5306, 5306 A, 5311, 4053); Shen-si, *Giraldi* (No. 128).

Numbers 5680, 5680 A, 5306, 5306 A, and 5311 may be considered the type of the species. The other specimens quoted differ somewhat in their shorter and broader more obovate leaves, more pubescent above, and are present only in fruiting specimens. *Lonicera saccata* is most nearly related to *Lonicera Tangutica*, Maximowicz, and its allies, differing from them chiefly by the prominent saccose gibbosity at the base of the tube and the three-celled ovary. From *Lonicera obovata*, Royle, and *Lonicera ramosissima*, Franchet & Savatier, which it resembles in the shape of the corolla, it is readily distinguished by the absence of the bractlets, by the foliage, and by the three-celled ovary. From Dr. Henry's collection *Lonicera saccata* was distributed as *Lonicera microphylla*, Pallas, var., but that species differs widely in the more or less distinctly bilabiate corolla, the shape of the leaves, and the two-celled ovary.

Alfred Rehder.

Arnold Arboretum.

EXPLANATION OF THE PLATE.

PLATE XX. LONICERA SACCATA.

1. A flowering branch, natural size.
2. A pair of flowers, enlarged.
3. A corolla laid open, enlarged.
4. An ovary with style, enlarged.
5. Cross section of a pair of ovaries, enlarged.
6. Vertical section of an ovary, enlarged.
7. A fruiting branch, natural size.
8. A seed, enlarged.

C. E. Faxon del.

LONICERA SACCATA, Rehd.

LONICERA KŒHNEANA, Rehd.

Lonicera Kœhneana, *n. sp.*

Leaves petioled, ovate, obovate or rhombic-ovate, from 6 to 10 centimetres long, acuminate, rounded or narrowed at the base, above bright green and sparingly appressed pubescent, becoming almost glabrous at maturity, below grayish green or glaucescent and covered with soft grayish pubescence and reticulate at maturity, with the veins more or less dilated at their base; petioles hirsute, from 5 to 10 millimetres long. Flowers in peduncled pairs from the axils of the lower and middle leaves; peduncle from 1.5 to 2.5 centimetres long, erect, rather stout, pilose; bracts linear, pubescent, about as long as the ovary; bractlets suborbicular, half as long or as long as the ovary, long-ciliate and sparingly pubescent, enveloping the ovaries like a cupule, the pair of the same flower almost wholly connate; ovaries distinct, globose, from 1 to 2 millimetres in diameter, glandular and short-pubescent, three-celled; calyx about half as long as the ovary, cupulate, divided to the middle into five ovate obtuse pubescent and ciliate teeth; corolla bilabiate, about 1.5 centimetres long, yellow, densely appressed pubescent and somewhat glandular on the outside; tube short and wide, prominently gibbous at the base, villose within; limb about one and a half times as long as tube, the upper lips divided into four short broadly ovate lobes; stamens about as long as the corolla; filaments hirsute at the base; anthers linear-oblong on the under side with a longitudinal row of setose hairs; style pubescent to the apex, as long as the stamens. Fruit subglobose, apparently red.

An upright probably robust shrub with stout branches covered with gray somewhat shreddy bark, green branchlets more or less tinged with purple, and sparingly pilose, and rather large ovate pointed winter-buds covered by several pairs of imbricate ovate to ovate-lanceolate grayish pubescent scales.

China: Sze-chuen, *Henry* (Nos. 5613, 5894), Farges (No. 1204); Hupeh, *Henry* (No. 6052); Yunnan, *Delavay* (No. 444); Shen-si, *Giraldi* (Nos. 123, 124, 126).

Dr. Henry's specimens from Sze-chuen constitute the type of the species; the others differ but slightly in the size of the leaves or the pubescence of the flowers. *Lonicera Kœhneana* is most nearly related to *Lonicera chrysantha*, Turczaninow, and has hitherto been referred to that species. It differs from it, however, in the large connate bractlets, the pubescent globose and larger ovaries, the densely appressed pubescent corolla with a wide and strongly gibbous tube, the setosely pubescent anthers, and in the soft-pubescent under side of the leaves and their firmer texture. *Lonicera chrysantha* has much smaller ovate bractlets wholly distinct or only slightly connate at the base, ovoid small ovaries about 1 millimetre in diameter and only glandular, a more slender corolla with scattered spreading hairs on the outer surface, anthers glabrous or furnished with a few setose hairs at the base, and leaves which are pilose chiefly on the veins beneath.

This species is dedicated to Professor E. Koehne, the distinguished German botanist, whose arrangement of the cultivated species of Lonicera in his *Deutsche Dendrologie* is the best and most natural hitherto published.

Alfred Rehder.

Arnold Arboretum.

EXPLANATION OF THE PLATE.

PLATE XXI. LONICERA KŒHNEANA.

1. A flowering branch, natural size.
2. A corolla, enlarged.
3. A corolla laid open, enlarged.
4. An anther, enlarged.
5. A pair of ovaries with the style, enlarged.
6. Cross section of an ovary, enlarged.
7. A fruiting branch, natural size.

C. E. Faxon del.

LONICERA KOEHNEANA, Rehd.

LONICERA FERRUGINEA, Rehd.

Lonicera ferruginea, *n. sp.*

Leaves persistent, ovate to oblong-ovate or ovate-lanceolate, from 6 to 10 millimetres long, acuminate, cordate or subcordate at the base, coriaceous, on the upper surface dark green, glossy and almost glabrous except on the midrib, with impressed veins, on the under surface light green, reticulate, and covered with ferrugineous pubescence, particularly along the veins; petioles from 4 to 6 millimetres long, densely ferrugineous-hirsute. Flowers in axillary heads consisting of several short-stalked or almost sessile pairs of flowers and forming terminal panicles; peduncles of the lower heads from 1 to 2 centimetres long; bracts linear, densely hirsute, exceeding the calyx; bractlets ovate, about one half as long as the ovary, densely hirsute; calyx-teeth ovate-lanceolate, hirsute, somewhat shorter than the ovary; ovaries distinct, densely hirsute; corolla bilabiate, from 2.3 to 2.8 centimetres long, yellow, densely hirsute on the outside, the tube slender, sparingly hairy within, about one third longer than the limb, the upper lip divided from one third to one half of its length into four oblong lobes; stamens exceeding the limb; filaments sparingly hirsute below the middle; style glabrous, as long as the stamens. Fruit black, ovoid, crowned by the persistent calyx.

A twining shrub with slender branches; the young branchlets, petioles, peduncles, and nearly every part of the plant except the upper surface of the leaves covered with ferrugineous hirsute pubescence.

China: Yunnan, Szemeo forest, at 1700 metres altitude, *Henry* (No. 11921, 11921 B).

A fruiting specimen (Henry, No. 11921 A) collected in the same locality at a lower altitude differs from the type by its thinner oblong-ovate to oblong-lanceolate leaves and more slender peduncles. *Lonicera ferruginea* is most nearly related to *Lonicera macrantha*, De Candolle, differing from it chiefly by the corolla being only half as long, by the hirsute ovaries, and by its dense ferrugineous pubescence; from *Lonicera Giraldii* ined., to which it is similar in appearance, *Lonicera ferruginea* differs in the glabrous style, the hirsute ovary, and the more slender and somewhat longer corolla-tube.

Alfred Rehder.

Arnold Arboretum.

EXPLANATION OF THE PLATE.

PLATE XXII. LONICERA FERRUGINEA.

1. A flowering branch, natural size.
2. A corolla laid open, enlarged.
3. A pair of ovaries with the style, enlarged.
4. Cross section of an ovary.

C. E. Faxon del.

LONICERA FERRUGINEA, Rehd.

LONICERA ARIZONICA, Rehd.

Lonicera Arizonica, *n. sp.*

Lonicera ciliosa, Rose, *Contrib. U. S. Nat. Herb.* i. 121 (not Poiret) (1891).

Leaves ovate or sometimes oval, from 2 to 4.5 centimetres long, truncate or broadly cuneate at the base, obtuse, bright green above, glaucous beneath, ciliate, otherwise glabrous, or rarely somewhat villous-pubescent on the lower surface while young; petioles slender, from 6 to 12 millimetres long, sparingly pilose and glandular, sometimes furnished with stipular appendages at the base, only the uppermost pair of leaves on the flowering branches connate into an elliptic to oblong-elliptic disk, pointed at both ends, from 4 to 8 centimetres long, and somewhat constricted in the middle. Flowers appearing in June in terminal pedunculate one to three-whorled heads, the whole inflorescence glabrous; peduncles from 0.5 to 2.5 centimetres long; bractlets ovate, about one third as long as the ovary, the two subtending bracts of each whorl oblong, as long as the ovaries; calyx scarcely one third as long as the ovary, cupulate with five short triangular-ovate teeth; corolla yellow, more or less tinged with scarlet outside, from 3.5 to 4.5 centimetres long, narrowly tubular-funnelform, with a short five-lobed limb, the tube somewhat inflated or a little gibbous above the base, slightly contracted below the middle, glabrous on the outer surface, hairy within below the insertion of the stamens, the lobes somewhat unequal, the longest from 6 to 8 millimetres long; stamens about as long as the tube, with glabrous filaments inserted somewhat above the middle of the corolla; style glabrous, exserted.

A low bushy or sarmentose shrub, with slender often drooping branches covered with light brownish yellow shreddy bark, glabrous green young branchlets usually tinged with purple, and covered with a glaucous bloom.

Mountains of Arizona, from 2000 to 3000 metres above the sea-level: Rincon Mountains, June 19, 1884, *C. G. Pringle;* Willow Spring, June 16 to 18, 1890, *E. Palmer* (No. 537); Santa Rita Mountain, June 6, 1881, *C. G. Pringle* (No. 306); San Francisco Mountains, July 4, 1869, *E. Palmer;* June 23, 1891, *D. T. MacDougal;* July 15, 1892, *J. W. Toumey;* Mormon Lake, June 12, 1898, *D. T. MacDougal* (No. 104).

The specimens from the Rincon Mountains and Willow Spring constitute the type of the species; those from the Santa Rita Mountains and part of those from Mormon Lake differ in the more gibbous and less slender corolla and slightly exserted stamens, while those from the San Francisco Mountains and part of those from Mormon Lake have shorter petioles and the under surface of the leaves somewhat pubescent. *Lonicera Arizonica* is allied to *Lonicera pilosa*, Willdenow, and *Lonicera ciliosa*, Poiret, and hitherto has been usually referred to the latter species. From *Lonicera pilosa* it differs in the thin obtuse long-ciliate but otherwise usually glabrous leaves, the less gibbous corolla, and shorter bractlets; from *Lonicera ciliosa* by the slender scarlet corolla with almost equal lobes, the glabrous style, and the much smaller slender-petioled leaves. In shape and color of the corolla *Lonicera Arizonica* much resembles *Lonicera sempervirens*, Linnæus, but is readily distinguished from that species by the thin ciliate leaves and the short head-like inflorescence.

Alfred Rehder.

Arnold Arboretum.

EXPLANATION OF THE PLATE.

PLATE XXIII. LONICERA ARIZONICA.

1. A flowering branch, natural size.
2. A flower, natural size.
3. A flower of the form from the Santa Rita Mountains, natural size.
4. A corolla laid open, enlarged.
5. An ovary with style, enlarged.

C. E. Faxon del.

LONICERA ARIZONICA, Rehd.

LONICERA GRIFFITHII, Hook. f. & Thoms.

Lonicera Griffithii, Hooker f. & Thomson, *Jour. Linn. Soc.* ii. 173 (1858). — Aitchison, *Jour. Linn. Soc.* xviii. 64. — Buser, *Boissier Fl. Orient.* Suppl. 275.

Leaves petiolate, oblong to orbicular, from 3 to 5 centimetres long, obtuse, dark green above, light green below, the lower ones oblong to ovate-oblong, usually narrowed at the base, the middle ones ovate to orbicular-ovate, rounded or subcordate at the base and, like the lower ones, on slender petioles from 8 to 12 millimetres long, the uppermost pair usually orbicular, emarginate and subcordate, short-petiolate. Flowers appearing in July and August in terminal pedunculate heads consisting of two or three, rarely four, whorls of flowers; peduncle from 0.5 to 2 centimetres long, hirsute, and glandular; each whorl subtended by six ovate-lanceolate hirsute bracts, the two longer ones mostly exceeding the ovaries, the others half as long; the eight bractlets of each whorl connate into a common cupule glabrous or sparingly hairy and glandular, as high as the ovaries and enveloping them completely; ovaries distinct, hairy and glandular; calyx with five ovate ciliate teeth about one third as long as the ovary; corolla bilabiate, 2.5 centimetres long, white and flushed with rose color, hirsute and glandular on the outer surface, the tube glabrous within, slender, gradually narrowed toward the base, slightly shorter than the limb, the upper lip divided about one third into short oval lobes; stamens shorter than the limb; filaments glabrous; style slightly longer than stamens, with long spreading hairs about the middle.

A twining shrub, quite glabrous except the bracts, calyx-teeth, corolla and peduncle of the inflorescence, with branches covered with chestnut-brown bark peeling off in thin flakes and leaving the stems grayish white.

Afghanistan: without locality, *Griffith*; Kuram Valley, common from the base of Péwárkotal at an altitude of 2500 metres, to Alikhél and Káratígha, *Aitchison* (No. 535).

Lonicera Griffithii is most nearly related to *Lonicera Periclymenum*, Linnæus, from which it differs in the connate bractlets and the shape of the leaves. By the connate bractlets, a peculiarity which seems to have escaped hitherto unnoticed, *Lonicera Griffithii* differs from all other species of the section Periclymenum (§ Caprifolium DC.) and may constitute the subsection *Thoracianthæ* (from θωρακεῖον, a breastwork, because the cupule surrounds the flowers of each whorl like a breastwork).

Outside of the sections Periclymenum and Nintoa connate bractlets forming a cupule tightly inclosing the ovaries are known in several species, as in *Lonicera Iberica*, Bieberstein, *Lonicera cærulea*, Linnæus,[1] and some other less well-known species. Analogous to the mode of union in these species which have the middle flower of the cyme suppressed, in *Lonicera Griffithii* with three-flowered sessile cymes only the bractlets of the two lateral flowers unite and form the cupule, inclosing in *Lonicera Griffithii* also the middle flower, and connect at the same time with the bractlets of the opposite cyme; thus they form a cupule inclosing the whole whorl. The bracts of the middle flowers, which bear in their axils the two lateral flowers, remain free and appear together with the subtending bracts of the two opposite whorls as six free bracts below the cupule. The tendency of opposite leaves to unite, which shows itself in other species of the section Periclymenum by the union of the pairs of leaves below the inflorescence, seems to be transferred in this species to the group of transformed leaves immediately below the individual flowers.

Lonicera Griffithii would probably prove to be a handsome ornamental climbing shrub if introduced into cultivation, for Aitchison speaks of it as "a magnificent climber, with very handsome rose-colored flowers." In the northern

[1] *Lonicera cærulea*, which is described even in the most recent manuals as having connate ovaries and no bractlets, has really distinct ovaries covered tightly by a cupule of connate bractlets. This cupule grows with the ovaries and affects at maturity the appearance of a coating of a single fruit similar to that of the connate berries of *Lonicera oblongifolia*, Hooker. So far as I know this fact was first pointed out by E. Koehne in his *Deutsche Dendrologie*, pp. 542, 545 (1893), and was also shown two years earlier in his drawings of Lonicera reproduced by K. Fritsch in Engler & Prantl, *Natürl. Pflanzenfam.* iv. pt. iv. 167.

United States, however, and in northern Europe it can hardly be expected to thrive without protection during the winter.

ALFRED REHDER.

Arnold Arboretum.

EXPLANATION OF THE PLATE.

PLATE XXIV. LONICERA GRIFFITHII.

1. A flowering branch, natural size.
2. Diagram of a whorl of flowers.
3. A corolla laid open, and half of the lowest whorl of the inflorescence, enlarged.
4. An ovary, enlarged.

C. E. Faxon del.

LONICERA GRIFFITHII, H. f. & T.

ENKIANTHUS SUBSESSILIS, Makino.

Enkianthus subsessilis, Makino, *Tokyo Bot. Mag.* viii. 215 (1894).

Andromeda subsessilis, Miquel, *Ann. Mus. Lugd. Bat.* i. 32 (1863–64); ii. 162; *Prol. Fl. Jap.* 94.—Maximowicz, *Bull. Acad. Sci. St. Pétersbourg*, xviii. 50; *Mel. Biol.* viii. 619.—Franchet & Savatier, *Enum. Pl. Jap.* i. 285.

Andromeda Nikoensis, Maximowicz, *Bull. Acad. Sci. St. Pétersbourg*, xxxi. 496 (1888); *Mél. Biol.* xii. 741.

Enkianthus Nikoensis, Makino, *Tokyo Bot. Mag.* viii. 215 (1894).—Palibin, *Scripta Botanica Hort. Petrop.* fasc. xv. 8 (*Revisio Gen. Enkianthus*).—Rehder, *Mitt. Deutsche Dendr. Gesell.* x. 113.

Leaves membranaceous, elliptic to rhombic-obovate, from 2 to 4.5 centimetres long, acute and apiculate at the apex, gradually narrowed below, and nearly sessile or borne on petioles from 1 to 5 millimetres long, crenulate-serrate, with incurved awned teeth, dark green on the upper surface, with short whitish hairs on the midrib, light green or glaucescent beneath, with rufous hairs along the midrib and on the veins, or only on the midrib. Flowers appearing in May or June with the leaves; racemes short-pedunculate, slender, nodding, from 4 to 6 centimetres long, from six to twelve-flowered, without bracts at the base; rachis villose; pedicels glabrous, slender, from 2 to 2.5 centimetres long, with a caducous linear bract at the base; calyx three times shorter than the corolla, divided almost to the base into ovate acuminate ciliate lobes; corolla globose-ovate, urceolate, white, 0.5 centimetres long, with a short five-lobed recurved limb, its lobes ovate-oblong; stamens about half as long as the corolla; filaments villose below the middle, dilated at the base; anthers glabrous, oblong with two awns at the apex as long as the cells; style glabrous, filiform, as long as the corolla; ovary five-celled. Capsules in nodding racemes, ovoid, 4 millimetres long, light brown; pedicels straight, thickened at the apex; valves furrowed at the back; seeds oblong, obtusely trigonous, 3 millimetres long, light brown, with a minutely reticulate testa.

A bushy shrub, from 1 to 3 metres in height, with irregularly whorled upright branches covered with gray smooth bark, and sparingly hairy or almost glabrous young branchlets.

Japan: Hondo, on the Nikko Mountains, *Savatier, Matsumura, Tschonoski* (ex *Palibin*), near Lake Chuzenzi, September 3 and November 3, 1892, *C. S. Sargent;* Province Shenano, Kinposan, at 1000 metres altitude, *Tschonoski* (ex *Palibin*).

Enkianthus subsessilis belongs to the section Andromedina of Palibin, distinguished by the urceolate corolla and the straight pedicels of the capsules. The only other species of this section is *Enkianthus Japonicus*, Hooker f., which is readily distinguished by the corolla having five gibbosities at the base and by the umbellate inflorescence with upright pedicels.

As an ornamental plant *Enkianthus subsessilis* is less handsome than the other species in cultivation, as the white flowers are very small, but, like that of the other species, the foliage assumes a brilliant red color in the autumn. It was introduced into the Arnold Arboretum by Professor Sargent, who collected seeds in Japan in 1892. It has proved hardy in the Arboretum, although it did not flower until 1901. I also saw it in bloom in the same year in Monsieur M. L. de Vilmorin's collection at Les Barres, France.

Alfred Rehder.

Arnold Arboretum.

EXPLANATION OF THE PLATE.

PLATE XXV. ENKIANTHUS SUBSESSILIS.

1. A flowering branch, natural size.
2. A flower, enlarged.
3. A stamen, enlarged.
4. A flower with the corolla and two sepals removed, enlarged.
5. A fruiting branch, natural size.
6. A fruit, enlarged.
7. Cross section of a fruit, enlarged.
8. A seed, enlarged.

C. E. Faxon del.

ENKIANTHUS SUBSESSILIS, Makino

The Riverside Press
Electrotyped and printed by H. O. Houghton & Co.
Cambridge, Mass., U. S. A.

ILLUSTRATIONS OF

LITTLE KNOWN LIGNEOUS P

REPARED CHIEFLY FROM MATERIAL AT
THE ARNOLD ARBORETUM OF
HARVARD UNIVERSITY

AND EDITED BY

CHARLES SPRAGUE SARGENT

Director of the Arnold Arboretum; Author of The Silva of North America

PART II

ATTERIA GRANDIFLORA
ATTERIA DOLICHOPODA
ATÆGUS REVERCHONI
ATÆGUS PALMERI
ATÆGUS DALLASIANA
CORNUS PURPUSI
CORNUS ARNOLDIANA
CORNUS BRACHYPODA
VIBURNUM SARGENTI
VIBURNUM VENOSUM

GUATTERIA GRANDIFLORA, DONN. SM.

GUATTERIA GRANDIFLORA, J. Donnell Smith, *Bot. Gazette*, xiv. 25 (1889).

Younger leaves linear-lanceolate, pilose beneath on the midrib; the older obovate or elliptical-oblong, from 12 to 20 centimetres long, from 4 to 7 centimetres broad, shortly cuspidate, cuneate at the base, coriaceous, glabrescent; petioles from 2 to 4 millimetres long, glabrous. Peduncles lateral from near the apex of the branchlets, solitary, from 1.5 to 2.5 centimetres long; flowers solitary, or usually from 2 to 6 in a corymbiform dichotomous cyme; bracts deciduous; pedicels from 1 to 3 centimetres long; sepals oblong-ovate, acute, 8 millimetres long, pubescent on both surfaces; petals subequal, oblong or elliptical-oblong, obtuse, from 3 to 4.5 centimetres long, velvety-pubescent, fleshy, the interior 3 tuberculate-papillose in a triangular space within near the base; apex of the torus concave; connective broadly dilated above the anthers, pulverulent; ovaries linear, with pubescent angles, twice longer than the style; stigma pulverulent. Berries nine to twelve, obtusely ellipsoid, 2.2 centimetres long, 1.2 centimetres thick, thrice longer than the stout stipe; seed rugulose, plicatures of the seed-coat very broad and intruding nearly to the centre of the albumen.

A tree, with glabrous branchlets and pubescent inflorescences. Guatemala: Department of Alta Verapaz; forests of Hacienda Pansamalá, at an altitude of 1250 metres, *Freiherr H. von Türckheim*, May 1887 (No. 1235 of my Series of Central American Plants); Hacienda Cubilquitz, at an altitude of 350 metres, *von Türckheim*, May, 1901 (No. 7816 of same Series).

This species is remarkable by its composite inflorescence, large flowers, and cleft albumen. It was originally described from imperfect material that exhibited only one-flowered peduncles.

Duplicates have been presented to the following Herbaria: Arnold Arboretum; Harvard University; Smithsonian Institution; Columbia University; Royal Bot. Gardens, Kew; British Museum; Königl. Bot. Museum, Berlin; Bot. Museum der Universität, Munich; Jardin des Plantes, Paris.

JOHN DONNELL SMITH.

Baltimore.

EXPLANATION OF THE PLATE.

PLATE XXVI. GUATTERIA GRANDIFLORA.

1. A flowering branch, natural size.
2. Posterior view of a flower, enlarged.
3. An interior petal, enlarged.
4. Vertical section of a torus, enlarged.
5. A stamen, front and rear views, enlarged.
6. A pistil, enlarged.
7. A cluster of berries, natural size.
8. A seed, enlarged.
9. Cross section of a seed, enlarged.
10. Vertical section of a seed, enlarged.

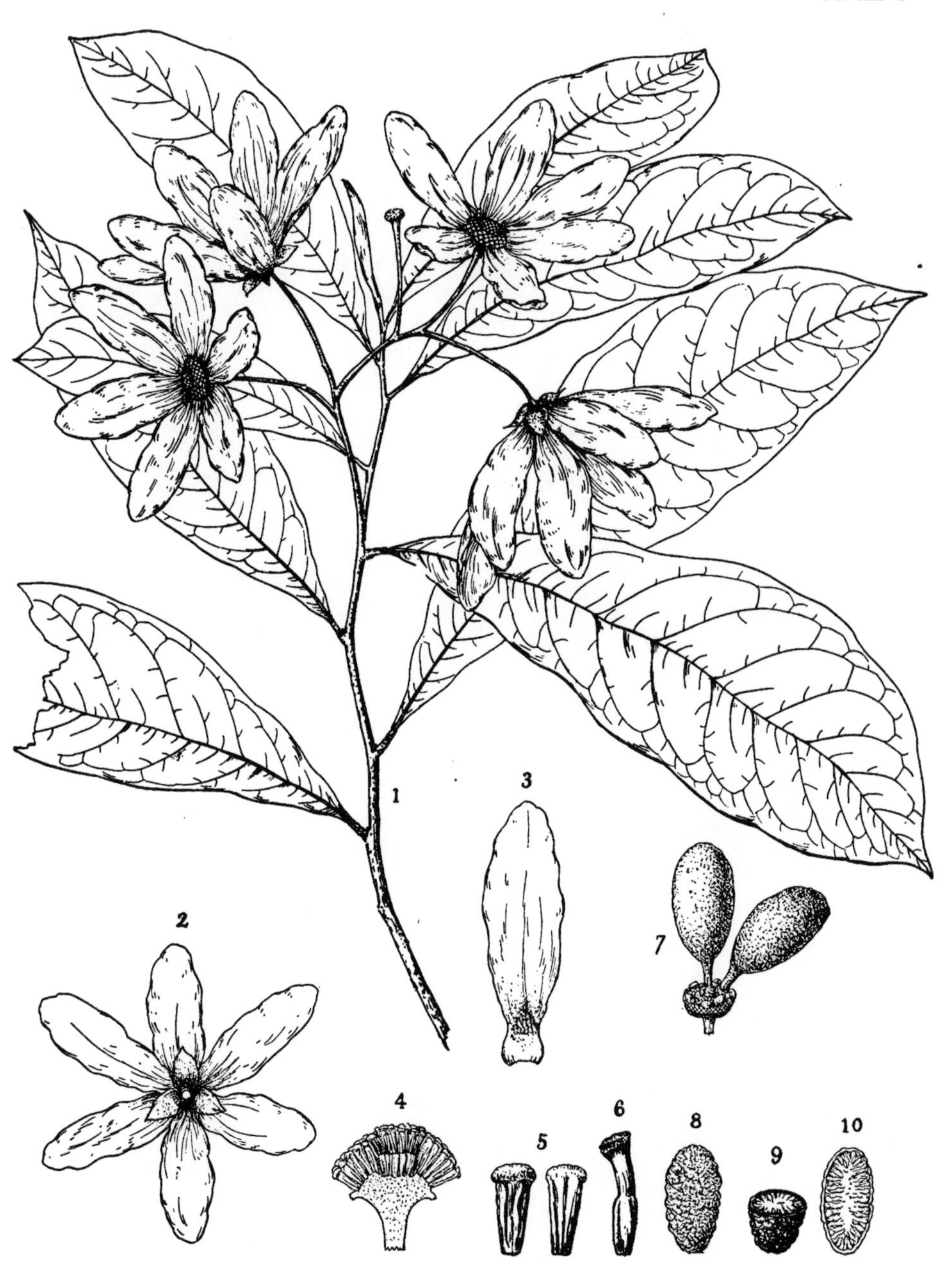

GUATTERIA GRANDIFLORA, Donn. Sm.

GUATTERIA DOLICHOPODA, DONN. SM.

GUATTERIA DOLICHOPODA, J. Donnell Smith, *Bot. Gazette*, xxiii. 2 (1897); *Pittier Primitiæ Floræ Costaricensis*, ii. 11.

Leaves lanceolate or elliptical-oblong, caudate-acuminate, acute at the base, from 12 to 18 centimetres long, from 3 to 5.5 centimetres broad, chartaceous, sparsely pilulose, at length glabrescent above; lateral nerves patent, veins minutely reticulate; petioles from 3 to 5 millimetres long, pilose. Peduncles axillary, solitary, one-flowered, from 3 to 4 centimetres long, pilose; bract at base and at the articulation below its middle oblong-ovate, from 2 to 4 millimetres long, glabrous above. Sepals ovate, acuminate, 5 millimetres long, glabrous within. Petals oblong, obtuse, somewhat unequal, from 1.9 to 2.2 centimetres long, from 6 to 9 millimetres broad, their lower half externally yellow-sericeous, elsewhere pubescent; apex of the torus excavate; ovaries linear, pilose; style short. Berries from thirty to fifty in number, green, ellipsoid, 8 millimetres long, 5 millimetres broad; stipes red, very slender, from 2.3 to 3.2 centimetres long; seed punctate, albumen penetrated by numerous linear protrusions of the seed-coat.

A tree, about 6 or 7 metres high, of irregular form, with sarmentose drooping branches, the ultimate ones clad with a dusky, patent pubescence; leaf-buds yellow-sericeous.

Costa Rica: Llanuras de Santa Clara, Hacienda La Concepción, at an altitude of 250 metres, *J. Donnell Smith*, February, 1896 (No. 6429 ex Pl. Guat. &c. quas ed. Donn. Sm.); forests of Xirores, Talamanca, at an altitude of 100 metres, *Pittier and Tonduz*, February, 1895 (No. 9166).

To this species are probably to be referred also the following numbers of the last named collectors, but the specimens seen exhibit neither developed flowers nor fruit: No. 10958 from Cañas Gordas, at an altitude of 1100 metres, March, 1897; No. 12970 from Rio De Las Vueltas, Tucurrique, on the Pacific slope, at an altitude of 635 metres, January, 1899.

Guatteria dolichopoda, as observed by me, occurs along rivers in the plains that lie at the foot of the great mountain-range of Costa Rica, some twenty miles back from the eastern coast. It attracts attention more by the abundant umbelliform clusters of fruit than by the flowers. They are conspicuous by the extremely numerous green berries contrasting with their coral-red filiform stalks.

Duplicates of No. 6429 may be found in the following Herbaria: Arnold Arboretum; Harvard University; Smithsonian Institution; Columbia University; Missouri Bot. Garden; Royal Bot. Gardens, Kew; British Museum; Königl. Bot. Museum, Berlin; Bot. Museum der Universität, Munich; Boissier Herbarium, Geneva.

JOHN DONNELL SMITH.

Baltimore.

EXPLANATION OF THE PLATE.

Plate XXVII. Guatteria dolichopoda.

1. A flowering branch, natural size.
2. A flower, posterior view, enlarged.
3. A flower, anterior view, enlarged.
4. Vertical section of a torus, enlarged.
5. A stamen, front and rear views, enlarged.
6. A pistil enlarged.
7. A fruiting branch, natural size.
8. A seed divided transversely, enlarged.
9. Vertical section of a seed, enlarged.

GUATTERIA DOLICHOPODA, Donn. Sm.

CRATÆGUS REVERCHONI, Sarg.

Cratægus Reverchoni, *n. sp.*

Leaves oval to obovate, acute, concave-cuneate and entire at the base, finely crenately serrate above, with gland-tipped teeth, nearly fully grown when the flowers open and then membranaceous, yellow-green and slightly villose near the base of the midribs on the upper surface, pale and glabrous on the lower surface; at maturity coriaceous, dark green and lustrous above, light yellow-green below, from 3 to 3.5 centimetres long, from 2 to 2.5 centimetres wide, with stout yellow midribs and five or six pairs of thin primary veins arching obliquely to the apex of the leaf; petioles stout, wing-margined nearly to the base, grooved, from 8 to 9 millimetres long; stipules linear, minute, turning red before falling, caducous; on leading shoots leaves usually nearly orbicular, rarely elliptical or ovate, coarsely serrate and occasionally slightly lobed, generally about 3.5 centimetres in diameter, with short broad-winged glandular petioles, bright red or rose color like the base of the thick midribs. Flowers from 8 to 9 millimetres in diameter on slender pedicels, in compact few-flowered thin-branched glabrous compound corymbs; calyx-tube narrowly obconic, the lobes slender, acuminate, tipped with small red glands, entire or slightly undulate or obscurely serrate on the margins; stamens ten to fifteen; anthers rose color; styles three to five, usually five. Fruit on slender elongated pedicels in drooping, few-fruited clusters, subglobose, often slightly broader than high, light scarlet, lustrous marked by occasional large dark dots, about 1 centimetre in diameter; calyx small, sessile, with a narrow deep cavity and elongated acuminate nearly entire lobes gradually narrowed from broad bases, mostly deciduous from the ripe fruit; flesh thick, light yellow, sweet and juicy; nutlets usually four, ridged on the back, with a broad rounded prominent ridge, from 6 to 7 millimetres in length.

A shrub, from 1 to 3 metres in height, with numerous stems rarely 1 decimetre in diameter and slender erect zigzag glabrous branchlets dull orange-brown when they first appear, soon becoming light red-brown and lustrous, and marked by numerous small pale lenticels, light red-brown in their second year and ultimately ashy gray, and armed with slender nearly straight red-brown or purple shining spines from 4 to 7 centimetres in length, often long persistent on the old stems and then occasionally from 1 to 1.5 decimetres long and furnished with numerous long lateral spines pointing upward. Flowers at the end of April or early in May. Fruit ripens the middle of September and then gradually falls during several weeks.

Upland prairies in rich black limestone soil, usually along the borders of woods near small streams, or occasionally isolated on the open prairie and then spreading into broad thickets. Dallas County, Texas: common; *J. Reverchon*, April, 1880, July, 1899, April and September, 1902.

I am glad to associate with this plant, which is one of the most distinct species of the Crus-galli Group, the name of the accomplished botanist and indefatigable collector, Jules Reverchon, who first made it known.

C. S. S.

EXPLANATION OF THE PLATE.

Plate XXVIII. Cratægus Reverchoni.

1. A fruiting branch, natural size.
2. Vertical section of a flower, enlarged.
3. A calyx-lobe, enlarged.
4. A fruiting branch, natural size.
5. Cross section of a fruit, enlarged.
6. Front view of a nutlet, enlarged.
7. Rear view of a nutlet, enlarged.
8. End of a terminal branchlet, natural size.

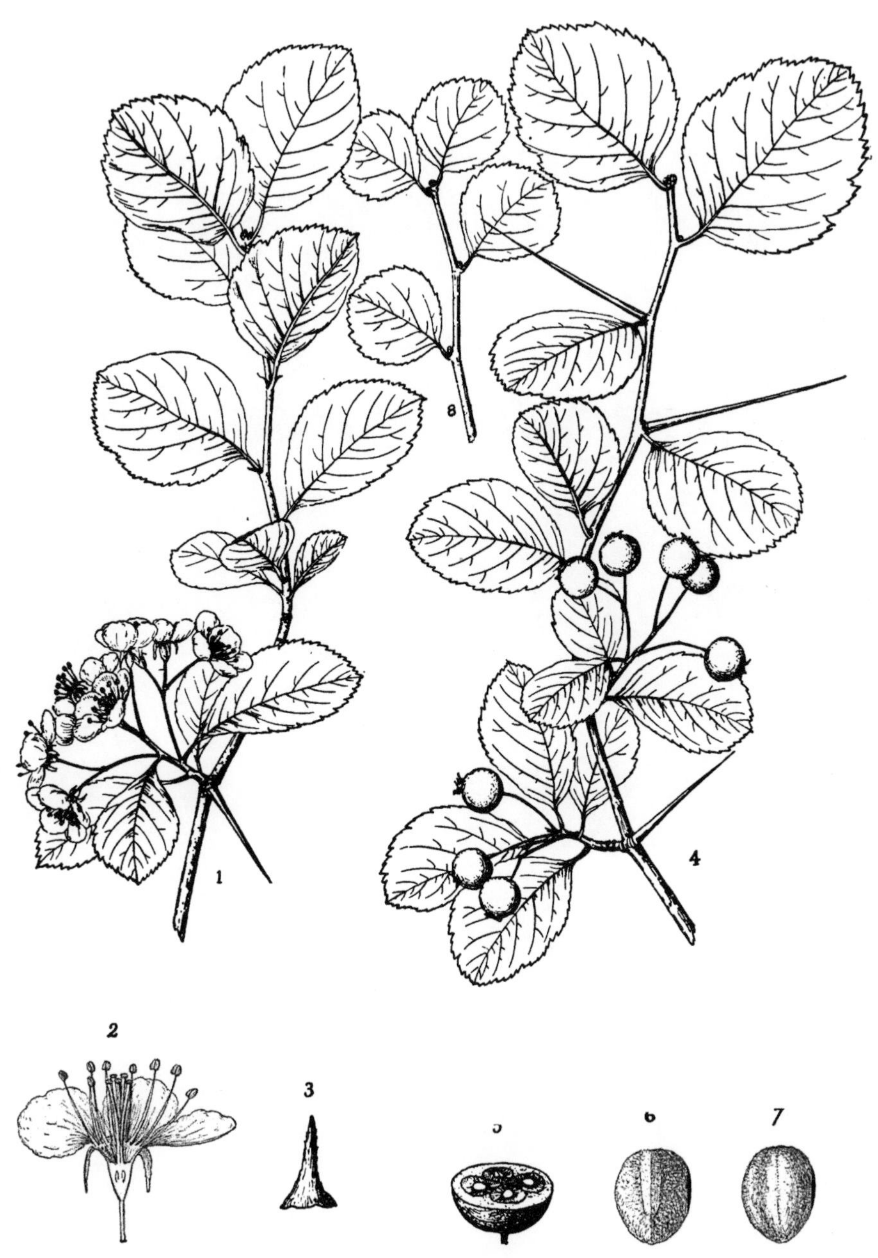

CRATÆGUS REVERCHONI, Sarg.

CRATÆGUS PALMERI, Sarg.

Crataegus Palmeri, *n. sp.*

Glabrous. Leaves broadly oval to obovate, rounded or acute or short-pointed at the apex, gradually narrowed and cuneate at the entire base, coarsely serrate above, with straight gland-tipped teeth; nearly fully grown when the flowers open and then very thin, dark green and lustrous above, pale bluish green below; at maturity coriaceous, dark green and lustrous on the upper surface, paler on the lower surface, from 4 to 6 centimetres long, from 3 to 4.5 centimetres wide, with slender yellow midribs and four or five pairs of very thin primary veins; petioles stout, slightly wing-margined toward the apex, grooved, often rose-colored in the autumn, about 1 centimetre in length; stipules linear, acuminate, often falcate, glandular, caducous; on vigorous shoots leaves oblong-ovate to elliptical, usually acute, coarsely serrate, occasionally laterally lobed, glandular at the base, with stipitate dark glands, from 6 to 7 centimetres long and from 4 to 5 centimetres wide. Flowers about 1.2 centimetres in diameter on slender pedicels, in compact many-flowered thin-branched compound corymbs; bracts and bractlets linear, minute, caducous; calyx-tube narrowly obconic, the lobes slender, acuminate, tipped with small dark glands, entire or slightly serrate; stamens ten; anthers pale yellow; styles three, surrounded at the base by a thin ring of pale tomentum. Fruit on slender elongated pedicels, in few-fruited drooping clusters, subglobose, bright cherry red marked by large pale dots, from 8 to 9 millimetres in diameter; calyx sessile, with a broad shallow cavity and erect and incurved lobes mostly persistent on the ripe fruit; flesh thin, yellow, dry and mealy; nutlets three, thin, acute at the ends, slightly and irregularly ridged on the back, with a low grooved ridge, from 6 to 7 millimetres in length.

A tree, sometimes 8 metres in height, with a trunk often 3 decimetres in diameter, covered with smooth pale bark, stout wide-spreading branches forming a broad round-topped symmetrical head, and slender nearly straight branchlets marked by few small pale lenticels, dark yellow-green when they first appear, soon becoming light red-brown and lustrous, dark reddish brown the following year and ultimately ashy gray, and armed with slender nearly straight dark red-brown shining spines from 2 to 4.5 centimetres long. Flowers during the first week of May. Fruit ripens the middle of October.

Missouri: Webb City, *E. J. Palmer*, May and October, 1901, 1902; Carthage, *Canby and Sargent*, May, 1902; *E. J. Palmer*, June and October, 1902; common, usually in low rich soil.

In the neighborhood of Carthage, where *Cratægus Palmeri* is abundant, the leaves are more often acute at the apex than on trees at Webb City, and the fruit is less brightly colored and frequently greenish when fully ripe. This handsome tree of the Crus-galli Group, one of the largest and most symmetrical of American Thorns, is named for its discoverer, Mr. E. J. Palmer, of Webb City, who has carefully collected and studied Cratægus in southwestern Missouri, where the genus is represented by a large number of interesting forms, of which several are still undescribed.

C. S. S.

EXPLANATION OF THE PLATE.

PLATE XXIX. CRATÆGUS PALMERI.

1. Part of a flowering branch, natural size.
2. Vertical section of a flower, enlarged.
3. A calyx-lobe, enlarged.
4. A fruiting branch, natural size.
5. Vertical section of a fruit, enlarged.
6. Cross section of a fruit, enlarged.
7. A nutlet, rear view, enlarged.
8. A nutlet, side view, enlarged.

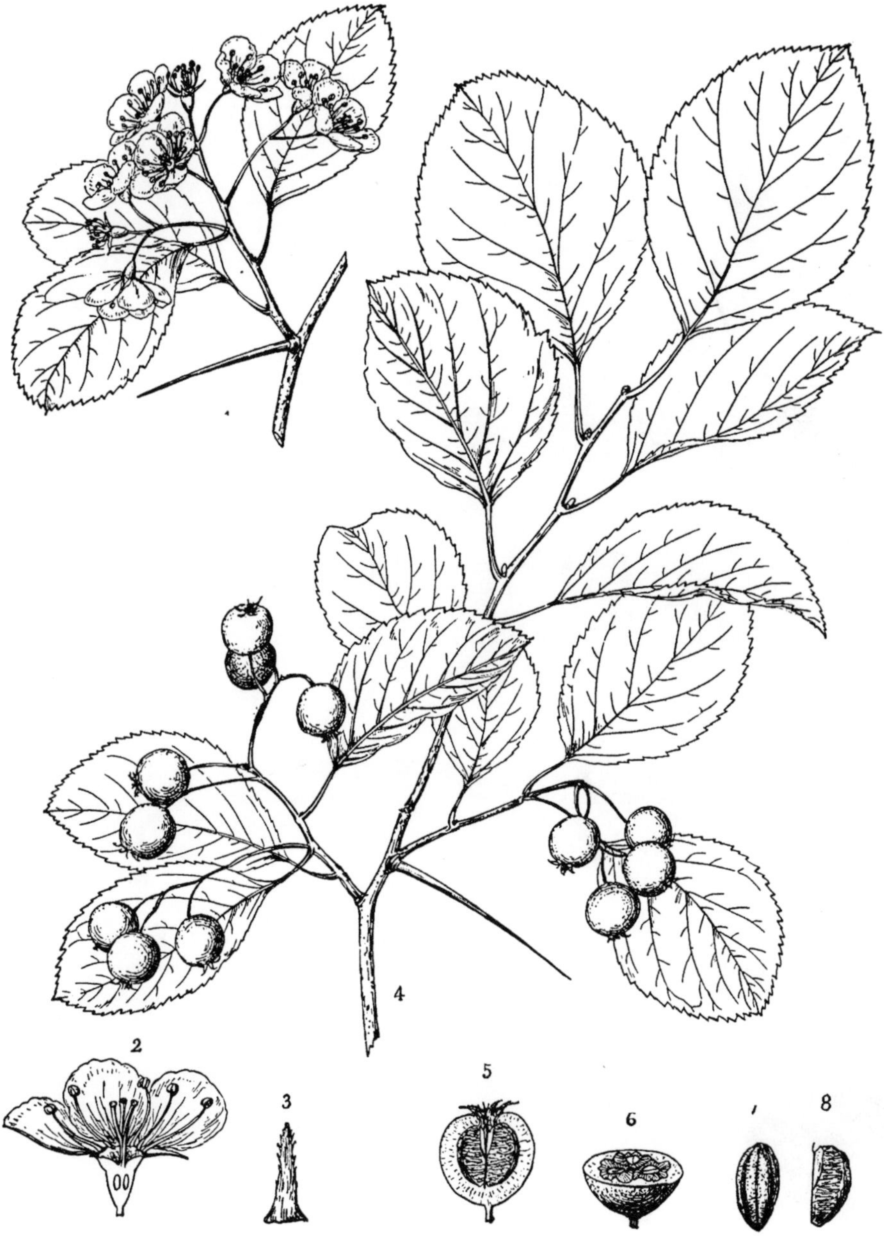

2 3 5 6 7 8 4

CRATÆGUS PALMERI, Sarg.

CRATÆGUS DALLASIANA, Sarg.

CRATÆGUS DALLASIANA, *n. sp.*

Leaves obovate, acute, acuminate or rounded at the apex, gradually narrowed to the concave-cuneate entire base, coarsely and double serrate above, with straight glandular teeth, and usually slightly lobed above the middle; coated below with thick hoary tomentum and villose above as they unfold, and nearly fully grown and villose or tomentose below when the flowers open; at maturity thin but firm in texture, dark yellow-green, glabrous and lustrous on the upper surface, pale and pubescent on the lower surface along the slender midribs and three or four pairs of thin arching primary veins, from 4.5 to 6 centimetres long, from 3 to 4 centimetres wide; petioles slender, wing-margined toward the apex, grooved, coated early in the season with hoary tomentum, glabrous in the autumn, from 1 to 1.2 centimetres in length; stipules linear, often falcate, small, caducous. Flowers about 1.6 centimetres in diameter, in many-flowered densely villose thin-branched compound corymbs; bracts and bractlets linear to narrowly obovate, turning rose color before falling, caducous; calyx narrowly obconic, densely coated like the slender pedicels with long matted pale hairs, the lobes slender, acuminate, tipped with minute red glands, sparingly and irregularly glandular-serrate, villose, reflexed after anthesis; stamens twenty; anthers light rose color; styles five. Fruit on stout erect slightly villose pedicels, in few-fruited clusters, subglobose, dark dull red; 9 or 10 millimetres in diameter; calyx persistent, with a short tube, a broad deep cavity, and spreading and reflexed lobes bright red on the upper side at the base; nutlets five, acute at the narrowed ends, thin, rounded and grooved, with a broad shallow groove or irregularly ridged on the back, about 7 millimetres long.

A tree, usually about 7 metres in height, with a tall trunk from 1 to 1.5 decimetres in diameter, covered with pale bark, small erect branches forming an open irregular head, and slender somewhat zigzag branchlets marked by many small pale lenticels, thickly coated at first with thick hoary tomentum, reddish brown and lustrous before autumn, lighter colored the following season and ultimately ashy gray, and armed with straight slender gray spines from 3 to 4 centimetres in length. Flowers early in April. Fruit ripens at midsummer.

In forest-covered bottom-lands of the small tributaries of the Trinity River, Dallas County, Texas; not common; *J. Reverchon*, April and July, 1900, April, 1901.

Closely related to *Cratægus Brazoria*, Sargent, of the Collina Group, and a native of the lower Brazos River, Texas, *Cratægus Dallasiana* differs from that species in the form and texture of the leaves, and in the dull red not canary-yellow early-ripening fruit, and in the color of the bark.

C. S. S.

EXPLANATION OF THE PLATE.

PLATE XXX. CRATÆGUS DALLASIANA.

1. A flowering branch, natural size.
2. Vertical section of a flower, enlarged.
3. A calyx-lobe, enlarged.
4. A fruiting branch, natural size.
5. Cross section of a fruit showing the nutlets, enlarged.
6. A nutlet, side view, enlarged.
7. A nutlet, rear view, enlarged.

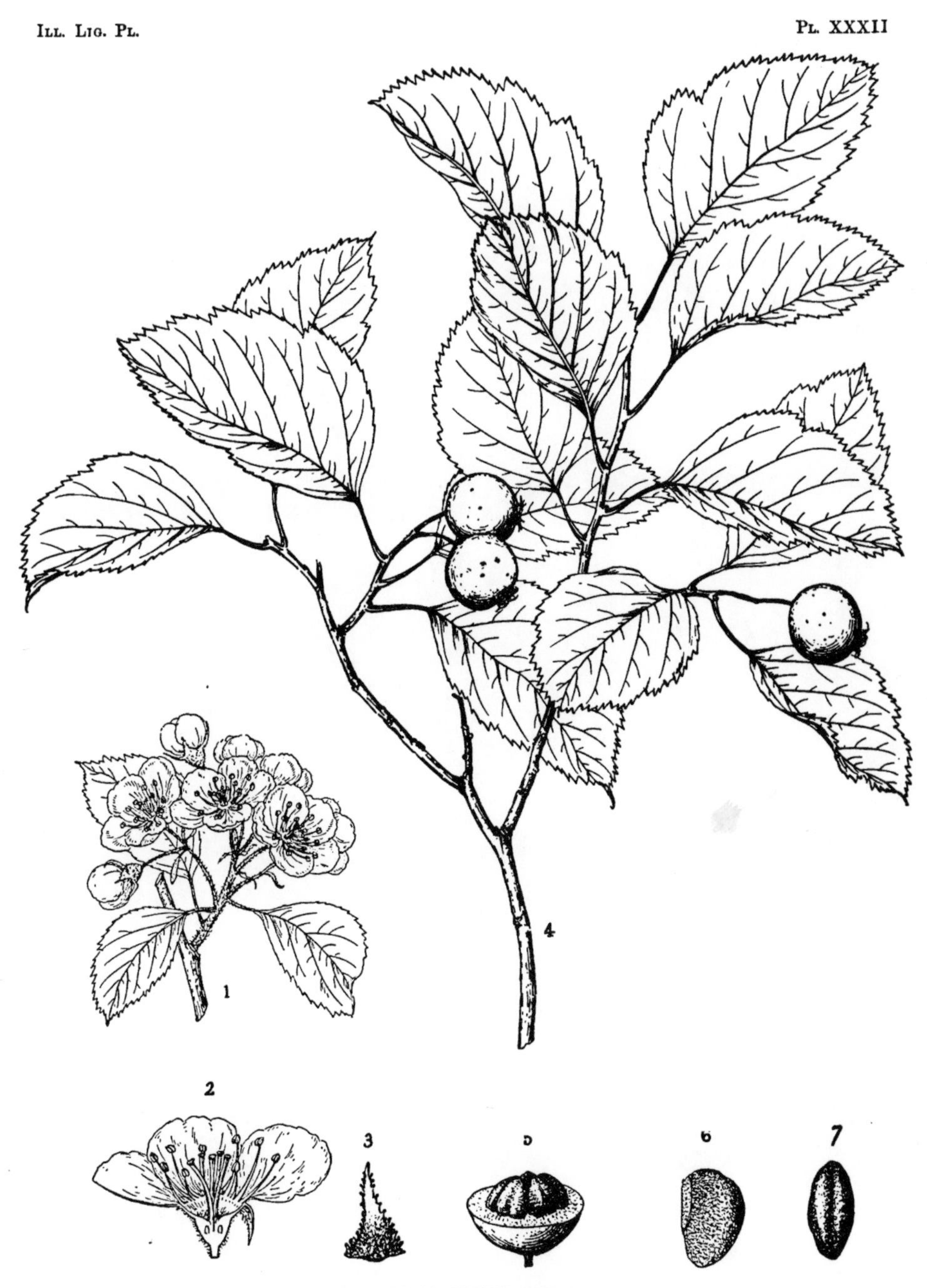

CRATÆGUS TRELEASEI, Sarg.

CRATÆGUS SPECIOSA, Sarg.

Cratægus speciosa, *n. sp.*

Leaves ovate, acute or acuminate, cordate, truncate or rarely rounded at the broad base, coarsely doubly serrate, with straight glandular teeth, and slightly divided into four or five pairs of small acuminate lateral lobes, tinged with red when they unfold and about one third grown when the flowers open, and then sparingly villose above and furnished below with persistent tufts of pale tomentum in the axils of the principal veins; at maturity thick and rigid in texture, dark yellow-green, glabrous and lustrous on the upper surface, pale on the lower surface, from 5.5 to 6 centimetres long, from 4 to 5.5 centimetres wide, with slender yellow midribs, and from five to seven pairs of thin primary veins arching obliquely to the points of the lobes; petioles stout, deeply grooved, slightly glandular, especially on vigorous shoots, often rose-colored in the autumn, from 1.4 to 2 centimetres in length; stipules oblong-obovate and acute to lanceolate, coarsely glandular-serrate, large, persistent like the much enlarged conspicuous inner bud-scales until after the flowers open. Flowers 2.5 centimetres in diameter on slender pedicels, in compact usually five to eight-flowered glabrous thick-branched compound corymbs; bracts and bractlets oblong, acute, frequently slightly falcate, glandular, conspicuous, persistent at least until May; calyx-tube broadly obconic, the lobes wide, acuminate, tipped with minute glands, coarsely glandular-serrate usually only above the middle; stamens twenty; anthers large, dark rose color; styles five, surrounded at the base by a broad ring of pale tomentum. Fruit in few-fruited drooping clusters, depressed-globose, full and rounded at the ends, obscurely five-angled, bright crimson, very lustrous, marked occasionally by large yellow-green blotches and by many small pale dots, about 2.2 centimetres broad and 1.9 centimetres high; calyx cavity 7 millimetres in diameter, the lobes only slightly enlarged, spreading and reflexed, mostly deciduous from the ripe fruit; flesh thin, acid, pale yellow; nutlets five, thin, acute at the ends, slightly grooved or irregularly ridged on the back, from 7 to 9 millimetres in length.

A shrub, with numerous stems from 2 to 3 metres in height, forming a broad open head, and stout nearly straight branchlets marked by small scattered pale lenticels, dark orange-green tinged with red when they first appear, bright red-brown and very lustrous during their first season and dull gray-brown the following year, and armed with numerous stout nearly straight purple shining spines from 2.5 to 5 centimetres long. Flowers during the last week of April. Fruit ripens and falls about the middle of September.

Dry hills, Prosperity Junction, near Webb City, Missouri; common; *B. F. Bush*, May 19, 1901; *E. J. Palmer*, September, 1901, April and July, 1902; *Trelease and Sargent*, October, 1901.

Cratægus speciosa, with its large flowers, large and brilliant fruits, and lustrous leaves, is a showy and beautiful plant and an interesting addition to the small Dilatata Group. It is most closely related to *Cratægus coccinioides*, Ashe, from which it differs in its longer and more slender pedicels, its larger flowers, smaller, early-ripening fruit with yellow not red flesh, in the much smaller calyx of the fruit with little enlarged spreading and appressed mostly deciduous lobes, smaller and more lustrous leaves, and in its shrubby habit.

C. S. S.

EXPLANATION OF THE PLATE.

PLATE XXXIII. CRATÆGUS SPECIOSA.

1. A flowering branch, natural size.
2. Vertical section of a flower, enlarged.
3. A calyx-lobe, enlarged.
4. A fruiting branch, natural size.
5. Vertical section of a fruit, enlarged.
6. Cross section of a fruit, enlarged.
7. A nutlet, rear view, enlarged.
8. A nutlet, side view, enlarged.

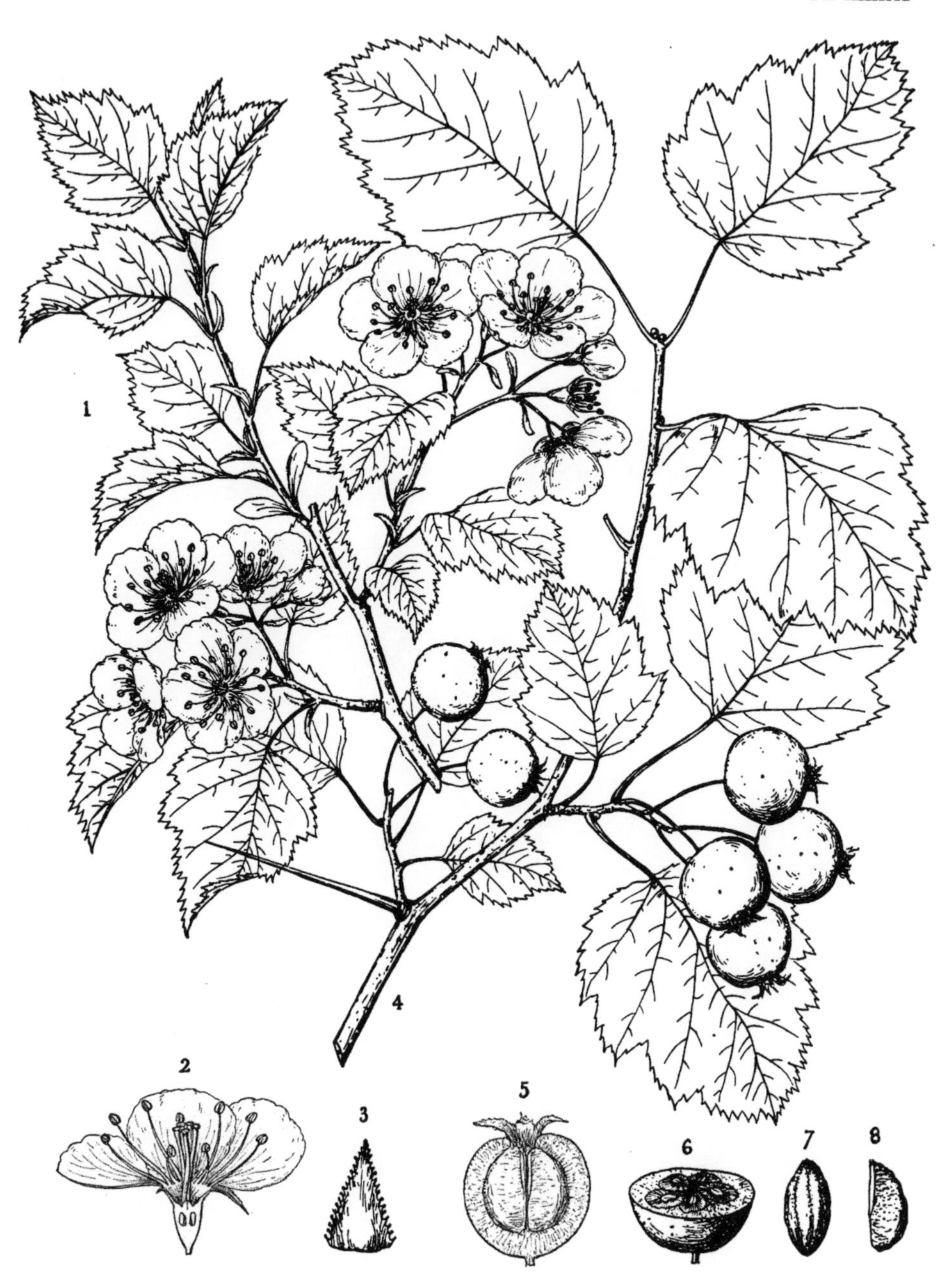

CRATÆGUS SPECIOSA, Sarg.

CRATÆGUS SMITHII, SARG.

CRATÆGUS SMITHII, *n. sp.*

Leaves obovate, rounded or acute at the apex, gradually narrowed from near the middle to the concave-cuneate entire glandular base, finely and doubly serrate above, with straight gland-tipped teeth, and occasioually divided into short terminal lobes; nearly fully grown when the flowers open and then membranaceous, slightly viscid, bright yellow-green and roughened above by short white hairs, paler, and villose below along the slender midribs and usually three pairs of thin primary veins extending to the apex of the leaf; at maturity subcoriaceous, very dark yellow-green, lustrous and scabrate on the upper surface, pale and still slightly hairy below, from 2.3 to 3 centimetres long, from 1.2 to 2.2 centimetres wide; petioles short, wing-margined nearly to the base, villose early in the season, pubescent in the autumn, from 4 to 5 millimetres long; stipules oblong and acuminate to lanceolate, glandular, turning brown in fading, caducous. Flowers solitary on short stout villose pedicels; bractlets linear to oblong, glandular, caducous; calyx-tube narrowly obconic, villose, the lobes foliaceous, broad-ovate, acuminate, conspicuously serrate, with slender teeth tipped with minute red glands, reflexed when the flowers open; stamens twenty; anthers pale yellow; styles five or rarely six. Fruit erect on short stout slightly hairy pedicels, short-oblong, full and rounded at the ends, orange or orange-green, lustrous, about 1.5 centimetres long and 1.2 centimetres wide; calyx enlarged, sessile, with a broad deep cavity and reflexed and closely appressed lobes; flesh thin, green and hard; nutlets five, rarely six, thick, gradually narrowed to the acute ends, irregularly ridged and deeply grooved on the back, 7 millimetres in length.

A dichotomously branched straggling shrub, from 1 to 2 metres in height, with slender nearly straight branchlets, orange-brown and coated when they appear with long pale matted hairs, dull reddish brown and pubescent at the end of their first season and dark gray-brown the following year, and armed with very slender straight dark purple spines, ashy gray and long persistent on the old stems, from 1.5 to 3 centimetres in length. Flowers about the 20th of May. Fruit ripens from the middle to the end of September.

Open hillsides along Lownes Run, Springfield, Delaware County, Pennsylvania, *George Smith*, May, 1867; *B. H. Smith*, May, 1902; *B. H. Smith, W. M. Canby*, and *C. S. Sargent*, September, 1902.

This interesting addition to the Uniflora Group may be distinguished from *Cratægus uniflora*, Moench, which grows near it and flowers from a week to ten days later, by the slender straight teeth of the leaves, the solitary flowers and much broader and shorter calyx-lobes, and the larger fruit. *Cratægus Smithii* in its name may help to perpetuate that of Dr. George Smith, one of the founders in 1833 of the Delaware Institute of Science, over which he presided for nearly half a century, an active promoter of the study of the natural sciences, and the author of *The History of Delaware County, Pennsylvania*, containing a notice of the geology of the county, and catalogues of its minerals, plants, quadrupeds, and birds. First collected by Dr. Smith in 1867, the specimen of *Cratægus Smithii* preserved in his herbarium led to its rediscovery more than thirty years later by his son, Mr. B. H. Smith, who continues enthusiastically and successfully the study of the flora of Delaware County.

C. S. S.

EXPLANATION OF THE PLATE.

PLATE XXXIV. CRATÆGUS SMITHII.

1. A flowering branch, natural size.
2. Vertical section of a flower, enlarged.
3. A calyx-lobe, enlarged.
4. A fruiting branch, natural size.
5. A nutlet, side view, enlarged.
6. A nutlet, rear view, enlarged.

CRATÆGUS SMITHII, Sarg.

CRATÆGUS MICRACANTHA, SARG.

CRATÆGUS MICRACANTHA, *n. sp.*

Leaves oblong-obovate to oval, acute, acuminate or rarely rounded at the apex, gradually or abruptly narrowed from above or below the middle to the cuneate entire base, coarsely crenulate-serrate and occasionally three-lobed above, the short lateral lobes broad and acute; when they unfold villose on the upper surface and coated below with thick hoary tomentum, and more than half grown when the flowers open and then membranaceous and slightly villose above, with short scattered pale hairs; at maturity thin but firm in texture, dark yellow-green, lustrous and smooth above, paler and villose or tomentose below along the slender midribs and three or four pairs of very obscure primary veins, from 5 to 6 centimetres long, from 2.5 to 3 centimetres wide; petioles slender, more or less wing-margined toward the apex by the decurrent base of the leaf-blades, tomentose early in the season, glabrous or pubescent in the autumn, from 1.5 to 2.5 centimetres in length; stipules linear, acuminate, elongated, caducous; on leading shoots leaves often broadly rhombic to obovate, acuminate, frequently deeply three-lobed or divided into two or three pairs of short lateral lobes, usually from 6 to 7 centimetres long. Flowers 6 millimetres in diameter on long slender pedicels, thickly coated with matted white hairs like the thin branches of the broad lax many-flowered compound corymbs; bracts and bractlets linear, elongated, turning brown in fading, mostly deciduous before the flowers open; calyx-tube narrowly obconic, villose, the lobes linear, acuminate, entire, slightly villose, tipped with minute dark glands, reflexed after anthesis; stamens usually ten, occasionally twelve, fifteen, or twenty; anthers very small, deep bright red; styles five. Fruit on slender pubescent pedicels, in drooping many-fruited clusters, subglobose to short-oblong, full and rounded at the ends, bright orange-red, lustrous, marked by occasional large pale dots, about 7 millimetres long; calyx prominent with a short villose tube and spreading erect hairy lobes often deciduous from the ripe fruit; flesh thin, yellow, dry and mealy; nutlets five, thin, acute at the narrowed ends, rounded and sometimes slightly grooved on the back, about 6 millimetres in length.

An unarmed tree, sometimes 8 metres in height, with a tall trunk from 3 to 4 decimetres in diameter, covered with light or dark brown bark separating freely into thin narrow scales which in falling disclose the light yellow-brown inner bark, stout spreading branches forming a broad flat-topped handsome head, and slender nearly straight branchlets coated until after the flowering time with thick white tomentum, bright red-brown marked by many large pale lenticels and puberulous at the end of their first season, becoming light or dark dull reddish brown the following year. Flowers at the end of April or early in May. Fruit ripens the middle of October.

Low woods in rich moist soil near Fulton, Arkansas, in the valley of the Red River; common; *B. F. Bush*, May 13, 1900, April and October, 1901, April, 1902; *Canby and Sargent*, April, 1901.

This species of the Viridis Group is distinct in the hairy tomentum which covers the young leaves, branches, and corymbs early in the season, in the small flowers, which are smaller than those of any other American species now known, the generally small number of stamens, and red anthers. The other species of this group are usually glabrous, with twenty stamens and light yellow anthers. *Cratægus micracantha* is one of the most interesting of the numerous Thorns discovered by Mr. Bush in southern Arkansas and Texas.

C. S. S.

EXPLANATION OF THE PLATE.

PLATE XXXV. CRATÆGUS MICRACANTHA.

1. Part of a flowering branch, natural size.
2. Vertical section of a flower, enlarged.
3. A calyx-lobe, enlarged.
4. A fruiting branch, natural size.
5. Vertical section of a fruit, enlarged.
6. Cross section of a fruit, enlarged.
7. A nutlet, side view, enlarged.
8. A nutlet, rear view, enlarged.

CRATÆGUS MICRACANTHA, Sarg.

MALUS SARGENTI, Rehd.

Malus Sargenti, *n. sp.*

Leaves slender-petiolate, ovate to elliptic or ovate-oblong, from 5 to 8 centimetres long, from 3 to 6 centimetres broad, acute or shortly acuminate, rounded or subcordate, rarely cuneate at the base, sharply and unequally serrate, with acuminate slightly spreading teeth; on flowering branches and short branchlets undivided and usually oval to ovate-oblong, on vigorous branches at least on the upper part ovate and three-lobed, and then usually subcordate at the base and shorter-petiolate, the middle lobe much larger than the others, broadest about the middle, often incised-serrate and even slightly lobed; covered as they unfold with white villose tomentum, soon disappearing from the upper surface; at maturity dark green and glabrous above, except on the impressed veins, pale green beneath and sparingly villose, particularly on the veins, the undivided leaves often nearly glabrous; petioles slender, from 1 to 2.5 centimetres long, pubescent; stipules, at least those of the upper part of the branches, leafy and persistent, ovate-lanceolate to lanceolate, often sparingly dentate. Flowers 2.5 centimetres in diameter, in broad five or six-flowered fascicles; pedicels slender, from 2.5 to 3 centimetres long, glabrous, upright or spreading; calyx-lobes like the calyx-tube glabrous on the outer surface, villose on the inner surface, ovate-lanceolate, acuminate, scarcely half as long as the petals; calyx deciduous before the maturity of the fruit; petals oval, from 1.2 to 1.5 centimetres long, overlapping, rounded at the apex, subcordate at the base and contracted into a short claw, pure white, glabrous; stamens fifteen to twenty, about half as long as the petals, with oval yellow anthers; styles usually four, sometimes five, rarely three, connate and villose below the middle; ovary four to five, rarely three-celled. Pome subglobose, about 1 centimetre in diameter, marked at the apex by the scar of the deciduous calyx, red, becoming soft and somewhat translucent when fully ripe; seeds obovate, about 4 millimetres long, light brown.

A low intricately branched shrub, about 1 or 1.5 metres high, with spreading rather rigid branches covered with smooth brown bark, and short often spinescent pubescent branchlets. Flowers appearing in May with the leaves. Fruit ripening at the end of September.

Japan: Hokkaido, brackish marsh near Mororan, September 25, 1892, *C. S. Sargent.*

Malus Sargenti is most nearly related to *Malus Toringo*, Siebold, differing from it chiefly in its larger pure white flowers, with broad subcordate petals overlapping each other, and in the larger fruits. From *Pyrus* (*Malus*) *Zumi*, Matsumura, which is also closely related to *Malus Toringo*, *Malus Sargenti* may be distinguished by the broader often lobed leaves, the shape of the petals, the glabrous calyx-tube, and the habit.

This handsome shrub was discovered by Professor Sargent during his travels in Japan, and was introduced by him into cultivation. Though naturally growing in marshy land, it thrives in drier localities at the Arnold Arboretum, where it has proved perfectly hardy and bears each year a profusion of pure white flowers, followed in autumn by attractive bright red fruits.

Alfred Rehder.

Arnold Arboretum.

EXPLANATION OF THE PLATE.

PLATE XXXVI. MALUS SARGENTI.

1. A flowering branch, natural size.
2. Vertical section of a flower with the petals removed, enlarged.
3. A petal, enlarged.
4. A stamen, enlarged.
5. Cross section of an ovary, enlarged.
6. A fruiting branch, natural size.
7. Vertical section of a fruit, enlarged.
8. A fruit, divided transversely, enlarged.
9. A seed, enlarged.

MALUS SARGENTI, Rehd.

ERIOLOBUS TSCHONOSKII, REHD.

ERIOLOBUS TSCHONOSKII, *n. comb.*

PYRUS TSCHONOSKII, Maximowicz, *Bull. Acad. Sci. St. Pétersbourg*, xix. 169 (1874); *Mél. Biol.* ix. 165. — Franchet & Savatier, *Enum. Pl. Jap.* ii. 349.

Leaves slender-petiolate, elliptic-ovate, from 7 to 12 centimetres long, acuminate, rounded or subcordate at the crenately serrate base, irregularly dentate-serrate, or oftener doubly serrate and even slightly lobed; when they unfold clothed with a floccose whitish tomentum on both sides, soon becoming glabrous on the upper surface, and dark yellowish green at maturity, floccose or villose-tomentose beneath, with six to ten pairs of lateral straight veins extending to the points of the teeth; petioles slender, subterete, from 2 to 3 centimetres long, floccose-tomentose when young, usually ultimately glabrous; stipules linear-lanceolate, about 1 centimetre in length, paleaceous, reddish brown, caducous. Flowers about 3 centimetres in diameter in two to four-flowered umbel-like racemes at the end of short branchlets, all or the lower flowers in the axils of foliage leaves; pedicels tomentose, from 2 to 2.5 centimetres long, furnished below the middle with two linear paleaceous caducous bracts from 8 to 10 millimetres in length; calyx-tube densely villose-tomentose; calyx-lobes triangular-ovate, 5 millimetres long, densely tomentose within, less tomentose without; petals white, elliptic-oblong, about 15 millimetres long and 8 millimetres broad, concave, obtuse at the apex, abruptly contracted at the base into short claws; stamens forty to fifty, the longest more than half as long as the petals; anthers ovate, yellow; styles five, not longer than the stamens, villose below the middle and connate for about one third of their length; ovary inferior, five-celled, the cells two-ovuled; carpels free at their ventral suture, leaving a central opening. Pome globose, about 3 centimetres in diameter, on a stout spreading or nodding pedicel, yellowish green with purple cheek, marked by numerous pale dots, crowned by the persistent connivent calyx-lobes; flesh firm, astringent, interspersed with small colonies of grit-cells, core portion free at the firm conical apex; seeds obovate, compressed, brown, about 8 millimetres long and 5 millimetres broad.

A large tree. Cultivated trees in the Arnold Arboretum ten years old are of narrow pyramidal habit with ascending or upright branches and spreading branchlets, sometimes spinescent on the older parts of the tree; young branchlets tomentose, becoming glabrous in the second year; bark of young branches dull brown, or dark gray on older branches, remaining smooth and marked by horizontal lenticels. Flowers appear about the middle of May shortly after the foliage. Fruit ripens the beginning of October and drops soon after maturity and at the time the leaves turn scarlet and orange.

Japan: Mountains of Central Hondo; foot of Fusi-san, *Tschonoski* (single fruits and leaves only, in herb. St. Petersburg!); Usui-toge near Kamisawa, *C. S. Sargent* (1892) (fruits only).

Eriolobus Tschonoskii appears to be a rare tree and to be confined to the central mountains of Hondo, where it probably grows only with scattered individuals. For a long time this interesting species was known only from a few single leaves and fruits picked up by Tschonoski from the ground, but the seeds collected by Professor Sargent in 1892 produced plants at the Arnold Arboretum which flowered two years ago and made it possible to ascertain the true characters of the species. Maximowicz, judging by the shape of the fruit, had referred it to Malus, but an examination of the flowers and of the fresh fruit shows that they resemble closely those of the section Cormus, which, however, differs in its compound inflorescence, except in the species referred by Decaisne[1] and Koehne[2] to Cormus as *Cormus trilobata*,

[1] *Nouv. Arch. Mus.* x. 157 (*Mém. Fam. Pomac.*). [2] *Gatt. Pomac.* 24.

which has a simple umbel-like raceme, and to this species *Pyrus Tschonoskii* is most closely related. As the inflorescence in Pomaceæ presents apparently constant and reliable characters, it seems desirable to separate the racemose *Cormus trilobata* from the other species and to place it with *Pyrus Tschonoskii* in the genus Eriolobus, Roemer (Pyrus, sect. Eriolobus, Seringe).

In regard to the position of this genus it may be observed that it differs from most Sorbeæ in its simple inflorescence, and of the genera with simple inflorescences it is closely related only to Docynia and Malus, while with Peraphyllum, Amelanchier, Chænomeles, Cydonia, and Pyrus it has little affinity. The chief character by which it differs from Malus is in the grit-cells of the fruit, while from Docynia it can be separated only by the number of the ovules. As the absence or presence of grit-cells seems to be a rather important character, and in this case is supported by other minor differences and by the general habit of the plant, a union of Eriolobus with Malus would hardly be natural, but Docynia is so similar in all its characters and in its general habit, that a separation by the three-ovuled carpels would appear rather artificial, as the number of ovules is not quite constant, and sometimes more than two ovules are found in some species of Malus and occasionally in other genera of Pomaceæ.[1] Docynia, therefore, will best be merged into Eriolobus, which, as thus enlarged, is apparently a natural genus, the species being very similar in their general habit and in the structure of their flowers and fruits.

The characters of the genus Eriolobus[2] as here proposed are as follows: Flowers in few-flowered umbel-like racemes, white; stamens numerous, with yellow anthers; styles five, united through their lower third; ovary distinctly inferior, five-celled, union of the carpels centripetalous, complete except at the free ventral suture, leaving a central space (or union complete without central space? according to Decaisne in his Docynia[3]). Pome rather large, with upright persistent calyx; flesh with grit-cells; the core portion free at the apex and attenuated into a firm conical point. Leaves conduplicate in the bud, doubly serrate or incisely lobed, glabrous or floccose-tomentose, deciduous.

In cultivation *Eriolobus Tschonoskii* grows into a handsome tree of narrow pyramidal habit, especially beautiful in autumn, when the foliage assumes brilliant orange and scarlet tints. It has proved perfectly hardy in the Arnold Arboretum, but has flowered and fruited hitherto only sparingly.

Alfred Rehder.

Arnold Arboretum.

1 Koehne, *Gatt. Pomac.* 25, 29.

2 *Eriolobus*, Roemer, *Syn. Mon. Rosifl.* 216 (1847).

Pyrus, § *Eriolobus*, Seringe, *De Candolle Prodr.* ii. 636 (1825).

Cormus, § *Eriolobus*, Decaisne, *Nouv. Arch. Mus.* x. 157 (*Mém. Fam. Pomac.*) (1874). — Koehne, *l. c.* 24.

Docynia, Decaisne, *l. c.* 131 (1874). — Hooker, *Fl. Brit. Ind.* ii. 369. — Fritsch, *Engler & Prantl Pflanzenfam.* Teil iii. Abt. iii. 22. — Koehne, *l. c.* 24.

The genus as thus extended comprises four species, *Eriolobus trilobata*, Roemer, *l. c.* (*Cratægus trilobata*, Labillardière, *Icon. Pl. Syr.* iv. 15, t. 10 (1812). — *Eriolobus Tschonoskii*, Rehder. — *Eriolobus Indica* (*Pyrus Indica*, Wallich, *Pl. As. Rar.* ii. 56, t. 173 (1831); *Docynia Indica*, Decaisne, *l. c.*). — *Eriolobus Hookeriana* (*Docynia Hookeriana*, Decaisne, *l. c.* t. 15), all Asiatic and native to Japan, the Himalayas, and Asia Minor.

The description given by Franchet and Savatier of their *Pyrus Tschonoskii*, var. *Hoggii* (*Enum. Pl. Jap.* ii. 349), is so incomplete that it is impossible to decide whether this form belongs in Eriolobus. It may be the same Pyrus which was collected afterwards in Hondo near Nikko by Professor Sargent and figured as *Pyrus Tschonoskii* (*Garden & Forest*, vii. 54, f. 9). This, however, is a true Pyrus, and probably an undescribed species indigenous to Japan.

3 Decaisne, *l. c.* t. 8, f. 3.

EXPLANATION OF THE PLATE.

Plate XXXVII. Eriolobus Tschonoskii.

1. A flowering branch, natural size.
2. Vertical section of a flower, enlarged.
3. A petal, enlarged.
4. A stamen, enlarged.
5. Cross section of an ovary, enlarged.
6. A fruiting branch, natural size.
7. Cross section of a fruit, natural size.
8. Vertical section of a fruit, natural size.
9. A seed, natural size.

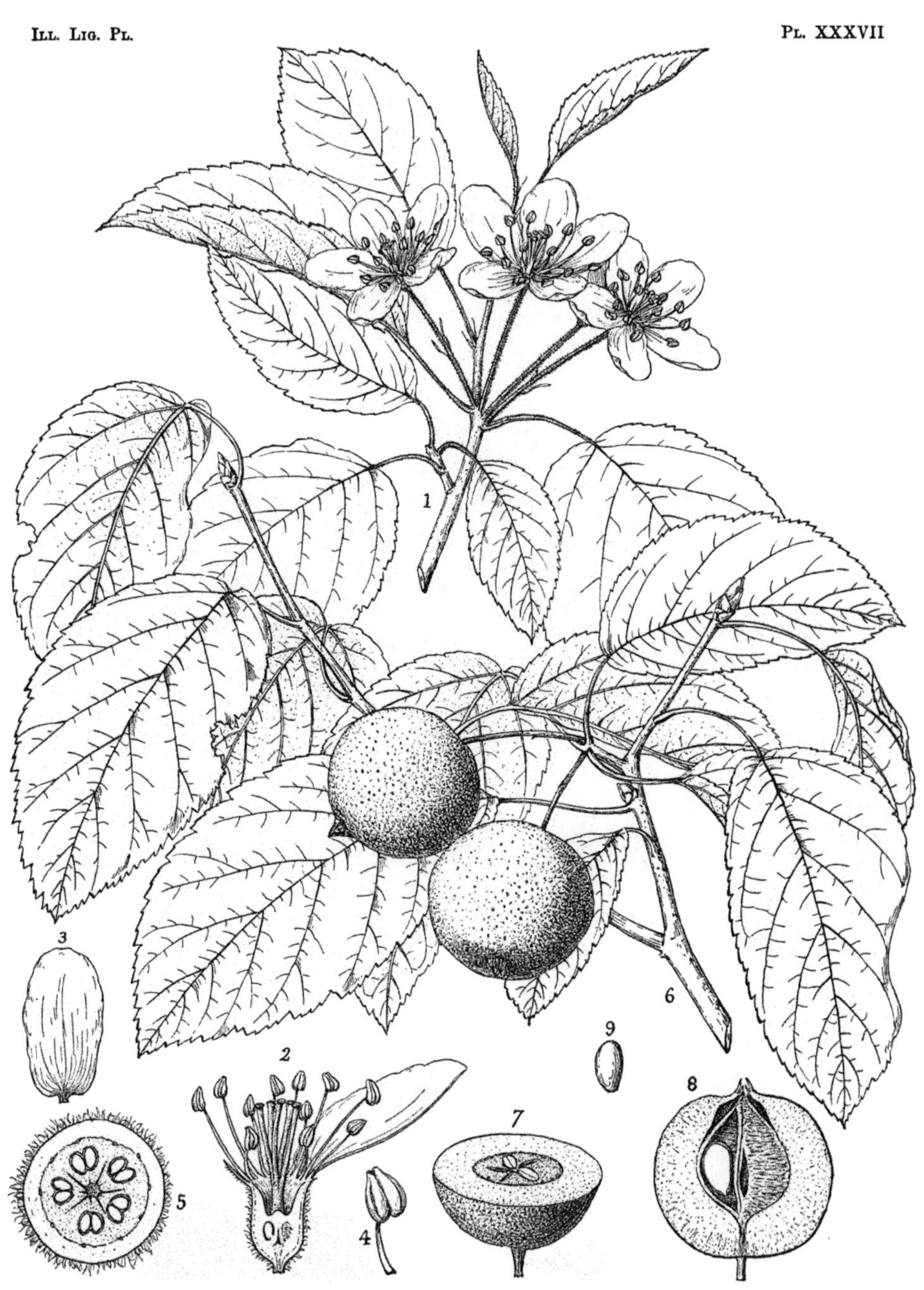

ERIOLOBUS TSCHONOSKII, Rehd.

RIBES FASCICULATUM, Sieb. & Zucc.

Ribes fasciculatum, Siebold & Zuccarini, *Abh. Akad. München*. iv. pt. ii. 189 (1845). — Miquel, *Ann. Mus. Lugd. Bat*. iii. 100; *Prol. Fl. Jap*. 264. — Maximowicz, *Bull. Acad. Sci. St. Pétersbourg*, xix. 264; *Mél. Biol*. ix. 237. — Forbes & Hemsley, *Jour. Linn. Soc*. xxiii. 279. — *Koehne, Deutsche Dendr*. 195. — Dippel, *Handb. Laubholzk*. iii. 302, f. 164. — Palibin, *Act. Hort. Petrop*. xvii. 91 (*Consp. Fl. Kor*.).

Leaves long-petiolate, alternate or fascicled, orbicular-ovate, three or rarely five-lobed, from 3 to 8 centimetres long, and about as broad, truncate or subcordate at the base; lobes ovate, acutish, incisely and coarsely serrate, serratures few, broad with convex margins, acutish; leaves dark green and glabrous on the upper side, pale green and glabrous beneath or puberulous on the veins when young, chartaceous at maturity; petioles from 1 to 2 centimetres long, puberulous above, dilated at the base, with several subulate ciliate appendages on each side. Flowers imperfectly diœcious, five-merous, glabrous, yellowish or greenish white, in few-flowered fascicle-like racemes appearing from lateral buds on branchlets of the previous year, together with fascicles of leaves, less often at the base of young branchlets; staminate flowers about 8 millimetres in diameter, in fascicles of three to five; pedicels slender, from 4 to 6 millimetres long, articulate about the middle, in the axils of short ovate bracts; calyx cupulate, with five spreading at last reflexed lobes, oval, 3 millimetres long; petals suborbicular, about one third as long as the sepals; stamens scarcely exceeding the petals, with the filaments shorter than the orbicular yellow anthers; style and ovary rudimentary; pistillate flowers in one to three-flowered fascicles, with shorter pedicels, a more campanulate calyx-tube, somewhat shorter sepals and rudimentary stamens; style shorter than the calyx-tube, with a two-lobed stigma; ovary ovoid, glabrous. Fruit a globose scarlet berry crowned by the persistent calyx-tube, about 1 centimetre in diameter, many-seeded; pedicels about as long as the fruit; seeds ovoid, about 4 millimetres long, sharply and irregularly three-angled, light pinkish brown.

An upright unarmed shrub, from 1 to 1.5 metres high, with short spreading branches, the younger covered with light yellowish gray shredding bark, the older with dark reddish brown flaky bark. Flowers appearing in May, followed at the end of September or at the beginning of October by scarlet berries of a sweetish but mawkish flavor. In Japan, according to Maximowicz, this shrub flowers in March and fruits in July.

Japan: Hondo, Nikko Mountains, *Tschonoski*, 1864; Tokio, *Miyabe; Yokohama*, *Maximowicz*, 1862; Shikoku, Nanokawa, *K. Watanabe*, March 29, 1890.

Ribes fasciculatum, which in its typical form is nearly or quite glabrous, is sometimes pubescent on the under surface of the leaves and on the petioles and young branchlets. This pubescent form, which has been described as var. *Chinense*,[1] occurs in northern China and Korea.

Both forms, the type and the pubescent variety, are in cultivation at the Arnold Arboretum, where they have proved perfectly hardy. The type was received in 1884 from the nursery of the S. B. Parsons Company, at Flushing, New York, under the name of *Ribes Japonicum;* the Chinese variety was introduced by Professor Sargent, who gathered ripe fruits in 1892 from plants cultivated in the Botanic Garden at Tokio. In the Arnold Arboretum the type is

1 *Ribes fasciculatum*, var. *Chinense*, Maximowicz, *Bull. Acad. Sci. St. Pétersbourg*, xix. 264; *Mél. Biol*. ix. 237 (1874). — Moore, *Jour. Bot*. xii. 138. — Ito, *Tokio Bot. Mag*. xiv. 105.

Ribes Chifuense, Hance, *Jour. Bot*. xiii. 36 (1875).

represented by staminate, the variety by pistillate plants. The bright scarlet fruits, which remain on the branches for a long time and do not begin to shrivel until the middle or end of November, make *Ribes fasciculatum* very attractive in the autumn, and a valuable addition to the shrubs with late-hanging ornamental fruits.[1] The leaves, too, do not fall until the beginning of the winter.

ALFRED REHDER.

Arnold Arboretum.

[1] Rehder, *Möller's Deutsche Gärtn.-Zeit.* xiv. 571 fig. (1899).

EXPLANATION OF THE PLATE.

PLATE XXXVIII. RIBES FASCICULATUM.

1. A branch with staminate flowers, natural size.
2. Vertical section of a staminate flower, enlarged.
3. Vertical section of a pistillate flower (var. *Chinense*), enlarged.
4. A sepal, enlarged.
5. A petal, enlarged.
6. A stamen, enlarged.
7. A fruiting branch (var. *Chinense*), natural size.
8. A seed (var. *Chinense*), enlarged.
9. A seed (var. *Chinense*), divided transversely, enlarged.

RIBES FASCICULATUM, Sieb. & Zucc.

CORNUS PURPUSI, Koehne.

Cornus Purpusi, Koehne, *Gartenflora*, xlviii. 338 (1899).

Leaves petiolate, elliptic-ovate to oblong, gradually or abruptly acuminate at the apex, cuneate or occasionally rounded at the base, on the upper side appressed-pubescent when unfolding, soon glabrous and dark green, on the under side glaucous (under the microscope covered with minute papillæ connected by ridges and folds of the epidermis), and sparingly appressed-pubescent on the five or six pairs of veins with whitish gray or fulvous appressed hairs, from 5 to 8 centimetres long, and from 2 to 4 centimetres broad; petioles slender, from 1 to 1.5 centimetres long, pubescent or sometimes nearly glabrous. Cymes dense, umbel-like, from 4 to 5 centimetres in diameter, with grayish or fulvous slightly appressed villose pubescence; peduncles from 3 to 5 centimetres long, pubescent. Flowers about 8 millimetres across, almost sessile or occasionally on pedicels as long or rarely longer than the ovary; calyx-teeth pubescent, ovate to subulate, one half or nearly as long as the densely appressed-pubescent ovary; petals oblong, acute; stamens slightly longer than the petals; style abruptly swollen below the stigma. Drupe subglobose, from 5 to 6 millimetres in diameter, dark blue or whitish and partly flushed with blue; stone irregularly subglobose, oblique, from 3.5 to 4 millimetres high, usually broader than high, irregularly and obtusely ridged.

A shrub, with spreading stems, spreading often arching branches, forming a loosely branched roundish bush, usually broader than high, and branchlets densely appressed-pubescent while young, becoming ultimately yellowish red or purple, in their second year slightly pubescent or glabrous, and finally changing toward the base of old stems to a dull gray. Flowers appear in July, or toward the southern limits of its range in May or June. Fruits ripen in September.

Northeastern North America from Quebec westward to Alberta, Minnesota, Nebraska, and Kansas, and southward to Missouri, Illinois, and Pennsylvania. Canada: Quebec, *J. G. Jack;* Alberta, *William M. Canby;* Maine, *J. C. Parlin;* New Hampshire, *M. L. Fernald;* Vermont, *C. G. Pringle, M. A. Day, A. Rehder;* Massachusetts, *L. L. Dame, A. Rehder;* Connecticut, *A. L. Winton;* New York, *A. Gray, A. Rehder;* Pennsylvania, *T. C. Porter;* Michigan, *E. J. Cole;* Wisconsin, *Mrs. Luce, J. G. Jack,* and *A. B. Seymour;* Ohio, *C. A. Purpus* (plants raised from seed collected near Toledo in the Botanic Garden, Darmstadt, Germany; type!); Illinois, *V. H. Chase;* Iowa, *R. Burgess, C. R. Ball;* Missouri, *B. F. Bush;* Kansas, *J. B. Norton;* Nebraska, *F. Clements.*

Cornus Purpusi is related to *Cornus Amomum*, Miller, and has been confounded until recently with that species. It differs from *Cornus Amomum* chiefly in the glaucous or glaucescent under side of the narrower leaves, which is densely covered by minute papillæ, while in *Cornus Amomum* the under side of the generally much broader leaves, which are usually rounded at the base, is pale or yellowish green, with a smooth epidermis. *Cornus Purpusi* has lighter colored yellowish red to purple branches; the pubescence of the inflorescence is more appressed; the inflorescence and the flowers are smaller, the pubescence on the under side of the veins of the leaves is less rusty, and sometimes is grayish white; the fruit is often pale, sometimes almost white, and the stone is less strongly ribbed. The whole appearance of the shrub with its looser habit and more or less pendulous leaves is quite distinct from the more compact *Cornus Amomum*. The geographical range of *Cornus Purpusi* is much wider than that of *Cornus Amomum*, which is restricted to the middle Atlantic States, ranging from Massachusetts southward to Georgia, and west to New York, Pennsylvania, and eastern Tennessee.

EXPLANATION OF THE PLATE.

PLATE XL. CORNUS ARNOLDIANA.

1. A flowering branch, natural size.
2. A flower, enlarged.
3. A pistil, enlarged.
4. A fruiting branch, natural size.
5. A stone, enlarged.

CORNUS PURPUSI, Koehne

CORNUS BRACHYPODA, C. A. MEY.

CORNUS BRACHYPODA, C. A. Meyer, *Ann. Sci. Nat.* sér. 3, iv. 74 (1845). — Walpers, *Ann.* ii. 725. — Miquel, *Ann. Mus. Lugd. Bat.* ii. 160; *Prol. Fl. Jap.* 92. — K. Koch, *Dendr.* i. 684. — Franchet & Savatier, *Enum. Pl. Jap.* i. 195. — Usteri, *Zeitschr. Gartenb. u. Gartenk.* xiv. 290. — Koehne, *Gartenflora*, xlvi. 94. — Makino, *Tokio Bot. Mag.* xi. 378. — Diels, *Engler Bot. Jahrb.* xxix. 506 (*Flora von Central-China*).

CORNUS SANGUINEA, Thunberg, *Fl. Jap.* 62 (not Linnæus) (1784). — Siebold & Zuccarini, *Abh. Akad. München.* iv. ii. 194.[1]

CORNUS ALBA, Thunberg, *Fl. Jap.* 63 (not Linnæus) (1784).[1]

CORNUS THELYCANIS, Lebas, *Rev. Hort.* 1875, 394, f. 64.

CORNUS CRISPULA, Hance, *Jour. Bot.* x. 216 (1881).

CORNUS IGNORATA, Shirasawa,[2] *Ic. Ess. Forest Jap.* i. 121, t. 77 (not K. Koch) (1900).

Leaves petiolate, opposite, membranaceous, elliptic-ovate to elliptic-oblong, from 7 to 14 centimetres long, from 4 to 7 centimetres broad, rather abruptly acuminate, rounded or broadly cuneate at the base, often finely erose-denticulate on the margins, with six to eight pairs of veins, dark green above and sparingly appressed-pubescent when young, becoming glabrous at maturity, glaucous beneath and sparingly beset with appressed forked hairs; petioles slender, grooved, from 1.5 to 2.5 centimetres long, glabrous. Cyme panicle-like and rather loose, from 8 to 15 centimetres in diameter, ramifications sparingly appressed-pubescent toward the tips; peduncle from 4 to 5 centimetres long, glabrous; pedicels shorter than the appressed-pubescent ovary or those of the terminal flowers about as long; calyx-teeth very short, triangular; petals oblong, acutish, from 4 to 5 millimetres long, appressed-pubescent on the outer surface; stamens slightly longer than the petals; style abruptly swollen below the stigma into a short disk. Drupe globose, from 5 to 6 millimetres in diameter, bluish black; stone subglobose, nearly smooth, about 4 millimetres in diameter.

A small tree or large shrub; young branchlets obscurely quadrangular, glabrous or nearly so, becoming yellowish or reddish brown the second year; bark of the trunk smooth, light gray. Flowers appearing in August. Fruits ripening in September.

Japan: Hondo, without locality, *Siebold;* Hakone, *Tschonoski*, 1864; Road Atami to Odawara, *C. S. Sargent*, August 28, 1892; Shikoku, Nanokawa, *K. Watanabe;* Kiu-siu, Nagasaki, *Oldham*, 1862; China: Prov. Hupeh, *A. Henry* (Nos. 5506, 6300 A), *E. H. Wilson* (Nos. 984, 1152, 1935).

Cornus brachypoda has often been confounded with *Cornus macrophylla*, Wallich, which, however, is very different in its alternate leaves and pitted stones. It is more closely related to *Cornus corynostylis*, Koehne, which is chiefly distinguished by its claviform style and smaller inflorescence. The *Cornus Theleryana*, Hort., which was doubtfully referred to *Cornus corynostylis* by Koehne, who had seen no flowering specimens, belongs, according to a flowering specimen from the Kew Arboretum, to *Cornus brachypoda*. *Cornus Theleryana* is like *Cornus Thelycanis*, Lebas, a name in all probability corrupted from *Cornus Thelycrania*, Thelycrania, Endlicher, being a section of the genus, and a shrub which I saw under the name *Cornus Thelycrania* in the Jardin des Plantes at Paris also belonged to *Cornus brachypoda*.

[1] Of *Cornus alba*, Thunberg, *Cornus sanguinea*, Thunberg, and *Cornus glauca*, Blume (ined.), there are specimens received from the herbarium at Leyden in the Gray Herbarium, which all belong to *Cornus brachypoda*.

[2] The figures of the fruits of *Cornus macrophylla* and *Cornus ignorata* have been apparently interchanged, for it is *Cornus macrophylla* which has the stone with a pit at the apex.

Cornus brachypoda is one of the handsomest Cornels in cultivation, if we except the species of the section Benthamia, and one of the most vigorous growing. Unfortunately it does not seem perfectly hardy in central Europe, though I noticed a good-sized shrub of this species in bloom in Späth's nursery near Berlin two years ago; farther south, in the public parks at Zurich, a tree of about 8 metres height, with a slender stem 38 centimetres in circumference, has been shown to me by Mr. A. Usteri, who had discovered it there several years ago, and in his paper quoted above correctly identified it with the much misunderstood *Cornus brachypoda* of C. A. Meyer. Its hardiness in the extreme northern United States is therefore doubtful; in the middle Atlantic States, however, it will certainly be hardy and a valuable addition to the list of ornamental trees.

ALFRED REHDER.

Arnold Arboretum.

EXPLANATION OF THE PLATE.

PLATE XLI. CORNUS BRACHYPODA.

1. A flowering branch, natural size.
2. A flower, enlarged.
3. Vertical section of a flower, enlarged.
4. A petal, enlarged.
5. A stigma, enlarged.
6. A fruiting cyme, natural size.
7. A stone, enlarged.
8. Cross section of a stone, enlarged.
9. Part of a leaf showing the under side with the forked hairs, enlarged.

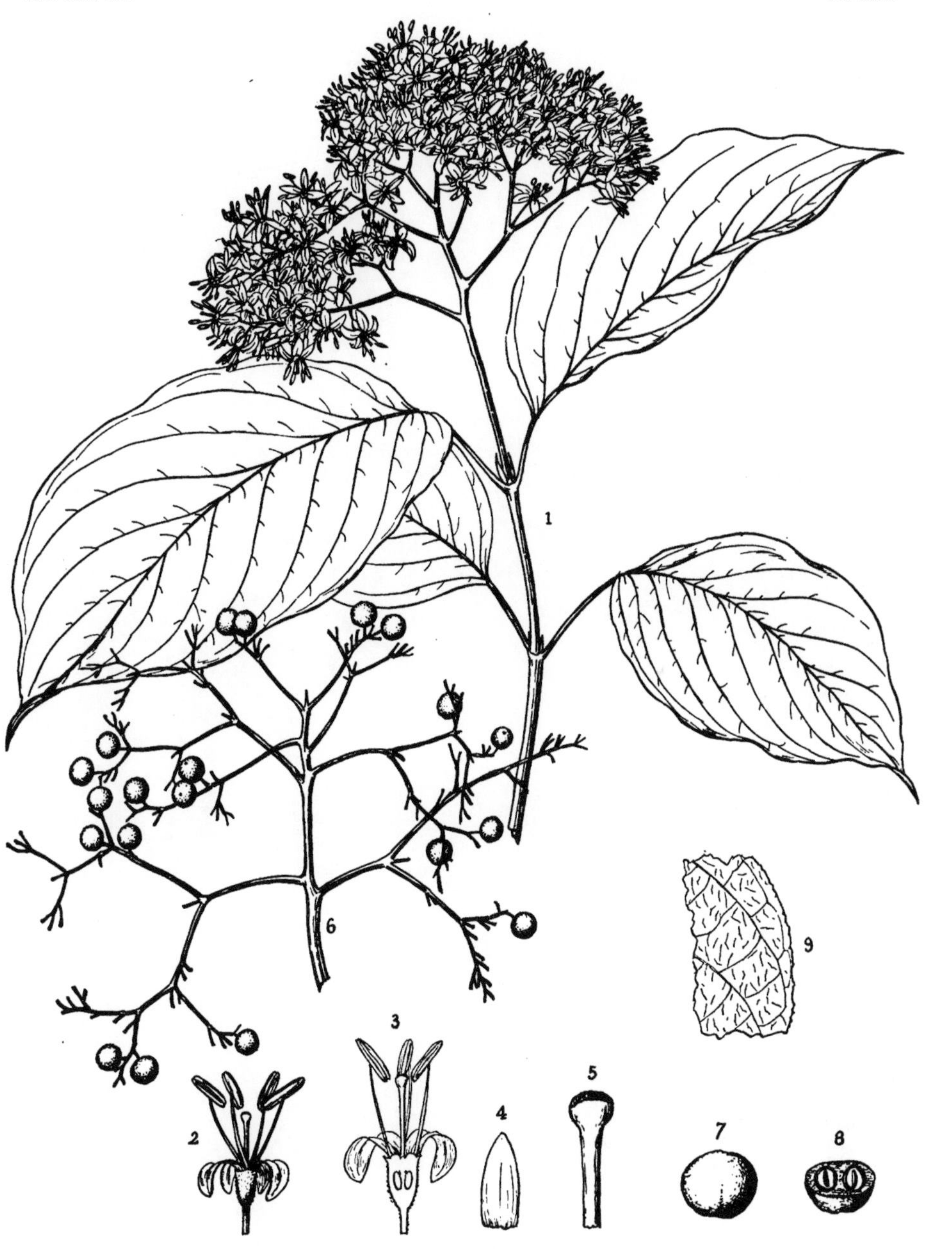

CORNUS BRACHYPODA, C. A. Mey.

VIBURNUM SARGENTI, Koehne.

VIBURNUM SARGENTI, Koehne, *Gartenflora*, xlviii. 341 (1899).—Rehder, *Bailey Cycl. Am. Hort.* iv. 1927.

VIBURNUM OPULUS, Gray, *Mem. Am. Acad.* n. ser. vi. 393 (*On the Botany of Japan*) (not Linnæus) (1859).—Miquel, *Ann. Mus. Lugd. Bat.* ii. 265; *Prol. Fl. Jap.* 153.—Franchet & Savatier, *Enum. Pl. Jap.* i. 199.—Maximowicz, *Bull. Acad. Sci. St. Pétersbourg*, xxvi. 4; *Mél. Biol.* x. 670.—Baker & S. Moore, *Jour. Linn. Soc.* xvii. 383.—Franchet, *Nouv. Arch. Mus.* sér. 2, v. 148 (*Pl. David.* i.).—Forbes & Hemsley, *Jour. Linn. Soc.* xxiii. 354.

Leaves petiolate, orbicular-ovate to ovate, from 6 to 12 centimetres long, from 5 to 10 centimetres broad, usually three-lobed and palmately three-nerved, rounded or truncate at the base, the lobes acuminate, coarsely and irregularly dentate, the lateral spreading; leaves on the upper part of the branches often elongated, narrowly elliptic to oblong-lanceolate, not lobed, remotely and sinuately dentate, or three-lobed, with entire or nearly entire acuminate lobes, the lateral lobes short, the middle lobe elongated; upper surface of the leaves dark yellowish green and glabrous, under surface pale green, short-pilose all over or only on the veins; petioles rather stout, from 2 to 4 centimetres long, pilose or nearly glabrous, grooved above, with two small subulate stipules at the base and two to four rather large disks at the apex. Flowers in umbel-like rather dense many-flowered six to eight-rayed corymbs from 8 to 10 centimetres in diameter, with radiant neutral flowers; peduncle stout, from 2 to 5 centimetres in length, pilose or glabrous; pedicels very short, like the minute distinctly toothed calyx, glabrous or nearly so; corolla cupulate-rotate, five-lobed, with short rounded lobes, about as long as the tube, creamy white, nearly 5 millimetres in diameter; stamens five, at least one and one half times as long as the corolla; anthers purple; stigma two-lobed, sessile; neutral flowers white, slender-pedicelled, from 2 to 2.5 or rarely to 3 centimetres in diameter, deeply five-cleft, their lobes obovate and unequal. Fruit subglobose, orange-scarlet, about 8 millimetres in diameter; stone orbicular, flattened, 6 millimetres in diameter, grayish white, slightly rough; embryo flat.

An upright shrub, from 2 to 3 metres in height, with branchlets pilose when young or glabrous, becoming yellowish or reddish brown in their second year and marked by conspicuous lenticels; bark of the stems and older branches dark gray, corky, with shallow longitudinal fissures; winter-buds ovoid, stipitate, of the color of the branches, glabrous, often partly covered with gum, inclosed by two outer connate scales. Flowers in June. The fruit ripens about the middle of September.[1]

Northern China; mountains near Pekin, *Bretschneider* (seeds only): Mandshuria, Ussuri, *Maack*; ad. fl. Amur, *Maack*, 1855, coast of Mandshuria, *C. E. Wilford*, 1859: Davuria, *Sosnin*: Japan, Hokkaido, Hakodadi, *C. Wright*, 1853–59, Mororan, *C. S. Sargent*, September 24, 1892, Sapporo, *Y. Tokubuchi*; Hondo, shores of Lake Chuzenji, *C. S. Sargent*, September 3, 1892.

The lower surface of the leaves, the branchlets and inflorescence of *Viburnum Sargenti* vary from pilose to glabrous. The glabrous form appears to be the more common one, as of the specimens referred to only those collected by Wilford,

1 *Viburnum pubinerve*, Blume in herb., is quoted by Miquel as a synonym of his *Viburnum Opulus*, and a specimen in the Gray Herbarium communicated from the herbarium at Leyden sustains this view, but according to Forbes & Hemsley (*Jour. Linn. Soc.* xxiii. 354) the specimen of *Viburnum pubinerve* in the Kew Herbarium belongs to *Viburnum phlebotrichum*, Siebold & Zuccarini.

EXPLANATION OF THE PLATE.

Plate XLIII. Viburnum venosum.

1. A flowering branch, natural size.
2. A flower, the corolla displayed, enlarged.
3. A fruiting branch, natural size.
4. A fruit divided transversely, enlarged.
5. A seed, enlarged.

VIBURNUM SARGENTI, Koehne

LONICERA MYRTILLUS, Hook. f. & Thoms.

Lonicera Myrtillus, Hooker f. & Thomson, *Jour. Linn. Soc.* ii. 168 (1858). — Brandis, *Forest Fl. Brit. Ind.* 255. — Buser, *Boissier Fl. Orient.* Suppl. 276. — Aitchison, *Jour. Linn. Soc.* xviii. 65. — Dippel, *Handb. Laubholzk.* i. 254, f. 166. — Koehne, *Deutsche Dendr.* 543.

Lonicera parvifolia, var. Myrtillus, Clarke, *Hooker Fl. Brit. Ind.* iii. 13 (1882).

Lonicera depressa, var. Myrtillus, Nicholson, *Hand-list Arb. Kew*, ii. 13 (1896).

Leaves short-petiolate, membranaceous, elliptic or elliptic-ovate to oblong, from 1 to 1.5 rarely to 2 centimetres long, obtuse, rounded or narrowed at the base, glabrous, dull green above, glaucescent beneath; petioles 1 to 2 millimetres long, glabrous. Flowers in short-peduncled pairs in the axils of the lower leaves; peduncles puberulous, about as long or sometimes longer than the petioles; bracts leafy, narrowly oblong, glabrous, exceeding the calyx; bractlets connate, with a truncate or slightly lobed cupule, about half as long as the connate ovaries; calyx cupulate, with five short triangular usually unequal teeth, glabrous; corolla cylindric-campanulate, yellowish white, glabrous without, about 1 centimetre long; tube with equal base, densely hairy within, especially near the mouth, about three times longer than the five-lobed spreading limb; lobes orbicular-ovate, obtuse; stamens inserted about the middle of the tube, with very short glabrous filaments; anthers not reaching the mouth; style glabrous, half as long as the tube, with a large capitate stigma; ovaries wholly or nearly connate, two-celled, the cells few-ovuled. Berries subglobose, about 6 millimetres in diameter, bright orange-red, glossy, crowned by the two persistent calyces; seeds few, ovoid, compressed, yellowish white, about 3 millimetres long, with a finely granulate testa.

A low much-branched shrub, from 0.5 to 1 metre in height, with slender upright or prostrate branches covered with light yellowish or grayish brown bark; young branchlets puberulous. Flowers appearing in May and June. Fruit ripens in July.

In the higher alpine regions of the Himalayan Mountains and Afghanistan, at altitudes of from 3000 to 4500 metres; Kashmir, Kishtwar, *T. Thomson*, without locality, *Jacquemont* (No. 2205); Punjab Himalaya, *Jaeschke;* Kumaon, *Strachey and Winterbottom;* Sikkim, *J. D. Hooker;* Afghanistan, *H. Collett*, Kuram Valley, *J. E. T. Aitchison* (No. 393).

Lonicera Myrtillus is not closely related to any other species with the exception of *Lonicera angustifolia*, Wallich, which, however, is easily distinguished from it by the larger acute or acuminate leaves, the slender drooping peduncles and the longer corolla-tube usually pubescent on the outside. The *Lonicera parvifolia* of Hooker f. & Thomson, which has been described as a closely allied species, can hardly be considered specifically distinct, as its distinguishing characters are slight and it is connected by intermediate forms with *Lonicera Myrtillus*. It is therefore referred as a variety to that species.[1]

[1] *Lonicera Myrtillus*, var. *depressa*, n. var.

Lonicera depressa, Royle, *Ill. Bot. Him.* 236 (1839), *nomen nudum.* — Nicholson, *Hand-list Kew Arb.* ii. 13.

Lonicera parvifolia, Hooker f. & Thomson, *Jour. Linn. Soc.* ii. 168 (1858) (not Hayne, nor Edgeworth). — Clarke, *Hooker Fl. Brit. Ind.* iii. 13.

This variety differs from the type in its larger and broader oval to oblong bracts and the longer peduncles sometimes almost as long as the leaf.

Himalayas: without locality, *Jacquemont* (No. 2237); Kumaon, *J. F. Duthie;* West Nepal, *J. F. Duthie;* Sikkim, *J. D. Hooker.*

The name *depressa* has been retained here for the *Lonicera parvifolia* of Hooker f. & Thomson, since it has been referred by most authors to this form, though Royle's type specimen in the Kew Herbarium, is of an intermediate character and might be

Lonicera Myrtillus was introduced into cultivation about twenty years ago and is growing at the Arnold Arboretum, where it has proved hardy with slight protection during the winter. It is a graceful little shrub, but the flowers are too small and, like the fruits, are not produced in sufficient numbers to make much ornamental effect.

ALFRED REHDER.

Arnold Arboretum.

referred as well to one as to the other. *Lonicera parvifolia* of Hooker f. & Thomson cannot be used as a name for this form on account of two older homonyms, for the *Lonicera parvifolia*, Edgeworth, erroneously referred by Hooker & Thomson to this form, is identical with the later *Lonicera obovata*, Royle, as Edgeworth's description (*Trans. Linn. Soc.* xx. 60 [1851]) and his type specimen in the Kew Herbarium clearly show. The name *Lonicera obovata*, however, cannot be replaced by *Lonicera parvifolia*, since the latter is homonymous to the still older *Lonicera parvifolia* of Hayne.

EXPLANATION OF THE PLATE.

PLATE XLIV. LONICERA MYRTILLUS.

1. A flowering branch, natural size.
2. A flower, enlarged.
3. Vertical section of a flower, enlarged.
4. A corolla laid open, enlarged.
5. Cross section of an ovary, enlarged.
6. A fruiting branch, natural size.
7. A seed, enlarged.

LONICERA MYRTILLUS, Hook. f. & Thom.

LONICERA THIBETICA, Bur. & Franch.

Lonicera Thibetica, Bureau & Franchet, *Jour. de Bot.* v. 48 (1891). — Mottet, *Rev. Hort.* 1902, 448, f. 198–200.

Leaves short-petiolate, deciduous, often in whorls of three, or rarely of four, oblong-lanceolate, from 1 to 3 centimetres long, acute or obtusish and mucronulate, rounded or abruptly narrowed at the base; on the upper surface dark green and minutely glandular when young, soon becoming glabrous and lustrous, densely white-tomentose on the lower surface; on sterile branches sometimes glabrous on both sides and dull bluish green above; petioles from 1 to 5 millimetres in length. Flowers in peduncled pairs in the axils of the middle and lower leaves; peduncles puberulous, from 1 to 10 millimetres long; bracts oblong to linear-oblong, acute or obtuse, about as long or sometimes longer than the calyx, sparingly glandular, tomentose beneath or nearly glabrous; bractlets nearly as long as the ovaries, rarely half as long, connate into a four-lobed cupule or sometimes (in cultivated plants) divided between the ovaries to the base, glandular-ciliate; calyx-teeth lanceolate, one third to one half as long as the corolla tube, glandular-ciliate, otherwise nearly glabrous; corolla pink or purplish pink, fragrant, salverform, from 1.2 to 1.6 centimetres long, rarely shorter, puberulous and glandular without; tube cylindric, equal at the base, furnished with long hairs within, especially near the mouth, from two to three times as long as the spreading five-lobed limb; lobes ovate, obtuse; stamens inserted about the middle of the tube; filaments very short, glabrous; anthers oblong, not reaching the mouth of the tube; style about half as long as the tube, glabrous; ovaries distinct, ovoid, glabrous, three-celled. Berries distinct, ovoid, about 1 centimetre long, orange-scarlet, glossy, crowned by the persistent calyx; seeds ovoid, compressed, about 5 millimetres in length, with a finely granulose testa.

A shrub, from 0.5 to 1.5 metres in height, with slender spreading and recurving often procumbent branches forming a dense intricate bush much broader than high; young branchlets villose-puberulous or tomentulose; older branches clothed with grayish brown shredding bark. Flowers appear in May and June and sparingly during the whole summer and autumn. Fruit ripens in August or September.

Tibet: between Lhassa and Batang, 1890, *Bonvalot and Henri d' Orléans* (type!); Kiala, *Soulié* (No. 238); China: Sze-chuen, west Sze-chuen and Tibet frontier, *A. E. Pratt* (No. 245), west Sze-chuen, *Przewalski, Potanin;* Ta-tsien-lu, *Soulié* (No. 261); Yunnan, *Delavay* (No. 2337).

Lonicera Thibetica is closely related to *Lonicera rupicola*, Hooker f. & Thomson, and to *Lonicera syringantha*, Maximowicz. From the former it differs in its narrower often acute leaves, their white tomentose under side and in the connate bractlets; from the latter chiefly in its pubescence, the shorter corolla-tube and the low prostrate habit.

Like those of *Lonicera rupicola*, the sterile branches and the leaves of *Lonicera Thibetica* often become perfectly glabrous, the dark glossy upper surface of the leaves changes to a dull bluish green color, and the leaves themselves often become broader and more obtuse, and then resemble those of *Lonicera syringantha*. Another peculiarity of the two species is a teratological tendency of cultivated plants toward dialysis and phyllody which shows itself in the calyx-teeth and the bractlets becoming more or less leaf-like, distinct, and elongated.

Lonicera Thibetica was introduced into cultivation in 1895, when seeds were sent from Sze-chuen by the Abbé Farges to Monsieur M. L. de Vilmorin of Paris.[1] In the Arnold Arboretum it has proved hardy with slight protection during

[1] *Jour. Soc. Nat. Hort. France*, sér. 3, xix. 743.

the winter and seems well suited for planting on rocky slopes and banks. The flowers are very fragrant, and the bright color of the berries makes the shrub ornamental in the autumn.

ALFRED REHDER.

Arnold Arboretum.

EXPLANATION OF THE PLATE.

PLATE XLV. LONICERA THIBETICA.

1. A flowering branch, natural size.
2. A pair of flowers of a cultivated plant, enlarged.
3. A pair of flowers of a specimen collected by Pratt in western Sze-chuen, enlarged.
4. A flower, the corolla displayed, enlarged.
5. Cross section of an ovary, enlarged.
6. A fruiting branch, natural size.
7. A seed, enlarged.

LONICERA THIBETICA, Bur. & Fr.

LONICERA TRAGOPHYLLA, Hemsl.

Lonicera tragophylla, Hemsley, *Jour. Linn. Soc.* xxiii. 367 (1888); Græbner, *Engler Bot. Jahrb.* xxix. 594 (*Flora von Central-China*).

Caprifolium tragophyllum, Kuntze, *Rev. Gen. Pl.* i. 274 (1891).

Leaves short-petiolate or sessile, ovate-oblong to oblong, from 7 to 14 centimetres long, membranaceous, with a narrow diaphanous margin, obtuse or subacute at the apex, cuneate at the base, dark green and glabrous above, glaucous and pubescent beneath, chiefly along the veins, rarely on the whole under side, the uppermost pair below the inflorescence connate into a suborbicular or oval disk, usually obtuse and mucronulate at both ends, the one or the two pairs below connate at the base. Flowers yellow, disposed in a terminal stalked head usually consisting of two whorls; peduncle glabrous, from 0.5 to 1.5 centimetres long; bractlets suborbicular, about one third as long as the ovate ovaries; calyx-teeth minute, triangular; corolla bilabiate, from 6 to 7.5 centimetres long, the tube slender, slightly curved, nearly three times as long as the limb, glabrous without, sparingly puberulous within, lower lip narrowly oblong, spreading, upper lip upright, divided for about one third its length into four oval or ovate rounded lobes; stamens nearly as long as the limb, glabrous, inserted at unequal heights, the one below the middle incision of the upper lip lower than the others; style glabrous, exceeding the stamens. Fruits scarlet, subglobose, about 1 centimetre in diameter, few-seeded; seeds ovate, compressed, 7 millimetres long, yellowish white, with a finely granulate testa.

A twining shrub, glabrous except the under side of the leaves; young branchlets green and usually partly tinged with purple; branches of the previous year light yellowish brown or grayish brown, smooth; winter-buds with several outer scales, ovate, acute, brown, and lustrous, the inner scales leafy.

China: Hupeh, *A. Henry* (Nos. 1707, 4010, 5898), *E. H. Wilson* (No. 1095); Sze-chuen, *Farges* (Nos. 109, 834, in herb. Paris), *A. von Rosthorn* (No. 1901 in herb. Berlin); Kansu, *Potanin* (in herb. Kew and Berlin).

Lonicera tragophylla seems most nearly related to *Lonicera Stabiana*, Pasquale, from which it differs in its much larger flowers and in the pubescence inside the tube; by the same characters and also by its shorter bractlets it differs from *Lonicera Etrusca*, Santi; from *Lonicera Caprifolium*, Linnæus, it is further removed by the stalked head-like inflorescence. It has the largest flowers of any species in the subgenus; and it is the only representative in central and eastern Asia of the almost exclusively Mediterranean subsection Eucaprifolium, Spach.

From its native habitat it is to be expected that this beautiful species may prove hardy in the United States and in central and perhaps northern Europe, at least with some protection during the winter.

Alfred Rehder.

Arnold Arboretum.

EXPLANATION OF THE PLATE.

PLATE XLVI. LONICERA TRAGOPHYLLA.

1. A flowering branch, natural size.
2. A corolla laid open, natural size.
3. A pistil, natural size.
4. A fruiting branch, natural size.

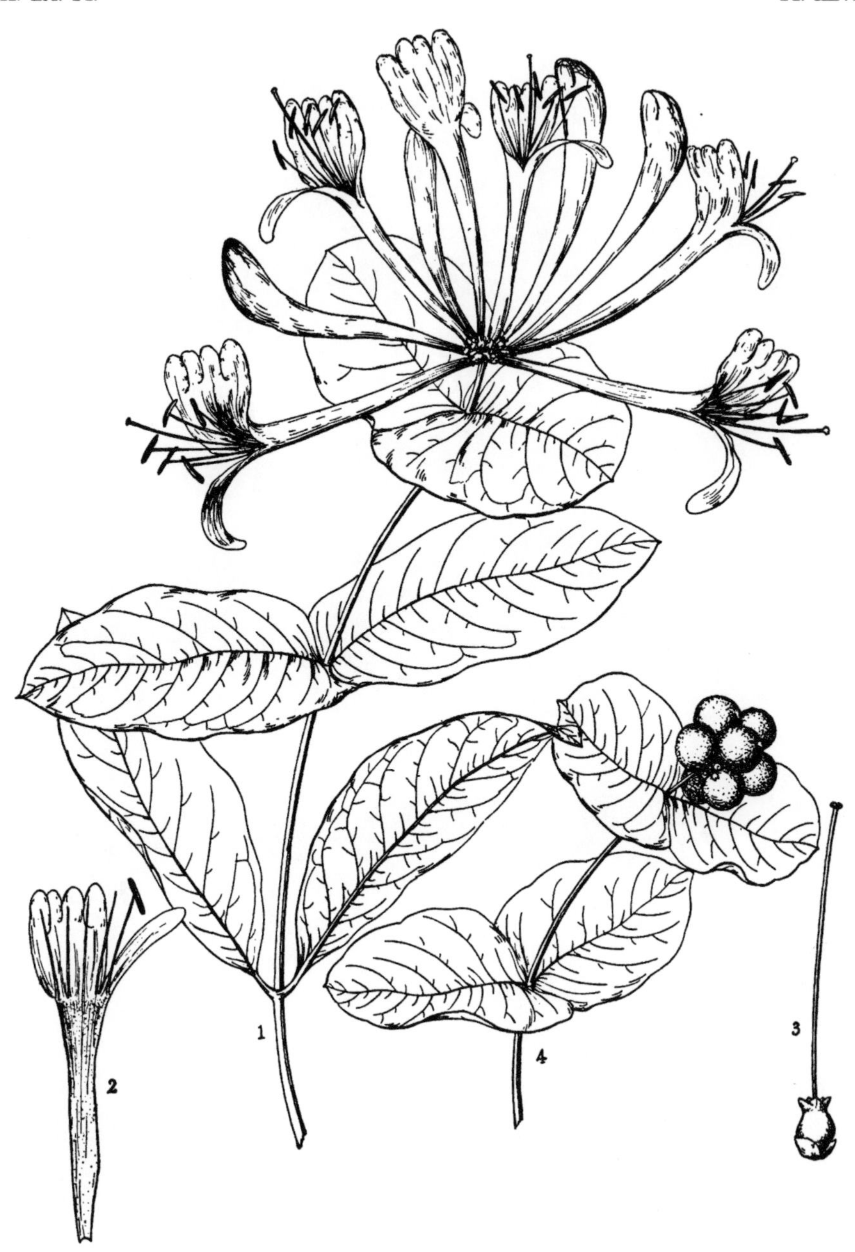

LONICERA TRAGOPHYLLA, Hemsl.

TECOMA HYBRIDA, Jouin.

(Tecoma radicans × Chinensis.)

Tecoma hybrida (Tecoma radicans × Chinensis), Jouin, *Jardin*, 1899 ex *Garden*, lv. 315 (1899).

Leaves impari-pinnate, from 15 to 30 centimetres in length; petioles glabrous, from 3 to 5 centimetres long; leaflets seven to eleven, ovate to elliptic-ovate, from 4 to 9 centimetres in length, coarsely serrate, usually long-acuminate, abruptly narrowed into petiolules rarely more than 1 centimetre long, the uppermost leaflets sometimes nearly sessile with the limb more or less decurrent, dark green, glabrous and somewhat lustrous on the upper surface, pale green below and usually pubescent along the veins, or rarely pubescent over the whole under surface. Flowers in terminal usually loose panicles often from 10 to 15 centimetres in length, with the lower branches from 2.5 to 5 centimetres long, becoming gradually shorter toward the apex, all terminated by three-flowered cymes; pedicels from 0.5 to 1 centimetre in length, with minute subulate deciduous bractlets; calyx campanulate, from 2 to 2.5 centimetres long, divided for about one third of its length into five ovate long-acuminate lobes; corolla funnel-form, campanulate, from 8 to 9 centimetres in length, with an orange-yellow tube and scarlet limb, the tube gradually widened from the tubular base into the spreading five-lobed limb about 5 centimetres in diameter, its lobes suborbicular; stamens four, two longer and two shorter; style somewhat shorter than the tube, surmounted by a two-lobed stigma. Fruit cylindric, from 15 to 20 centimetres long and 1.5 to 2.5 centimetres broad, gradually narrowed at both ends and usually curved, with winged sutures, light olive-brown; seeds compressed, light brown, about 1 centimetre long, with the hyaline wings oblong in outline and about 3 centimetres in length.

A shrub, climbing by aerial rootlets; young branchlets green and glabrous, older branches covered with pale yellowish white bark. Flowers from July to September. The fruits ripen in September or October.

The plant which is here figured and described was received at the Arnold Arboretum from the P. J. Berckmans Company of Augusta, Georgia, under the name *Tecoma hybrida*. It is obviously a hybrid between *Tecoma radicans*, Jussieu, and *Tecoma Chinensis*, K. Koch (*Tecoma grandiflora*, Loiseleur), as it is exactly intermediate between these two species in all its characters. From *Tecoma radicans* it is chiefly distinguished by the looser and larger inflorescence, the much wider corolla-tube with a broader limb, and by the longer acuminate calyx-teeth; from *Tecoma Chinensis* it differs in the pubescent covering of the under side of the leaflets, in the longer corolla-tube which much exceeds the calyx, and in the shorter calyx-lobes.

A plant, cultivated at the Arnold Arboretum as *Tecoma radicans grandiflora atropurpurea*, does not differ materially from this hybrid. According to Koehne,[1] *Tecoma Chinensis aurantiaca*, Hort., is probably a hybrid of the same parentage, and Jouin[2] believes *Tecoma Princei coccinea grandiflora*, Hort., to be also a hybrid between these two species.

As a garden plant *Tecoma hybrida* is valuable, for its flowers are almost as large and showy as those of *Tecoma Chinensis*, and it is hardier than that species. In the Arnold Arboretum, where the Chinese species cannot be grown in the open air, the hybrid lives through the winter when slightly protected. Like *Tecoma Chinensis*, it begins to flower while small, and it does not appear to climb so high as *Tecoma radicans*.

Alfred Rehder.

Arnold Arboretum.

[1] *Deutsche Dendr.* 522.

[2] *Gard.* lv. 315.

EXPLANATION OF THE PLATE.

PLATE XLVII. TECOMA HYBRIDA.

1. An inflorescence, natural size.
2. A flower, natural size.
3. A calyx with the style, natural size.
4. A leaf, natural size.
5. A fruit, natural size.
6. A flower of *Tecoma radicans*, natural size.
7. A calyx and style of *Tecoma radicans*, natural size.
8. A flower of *Tecoma Chinensis*, natural size.
9. A calyx and style of *Tecoma Chinensis*, natural size.

TECOMA HYBRIDA, Jouin

PICEA MORINDOIDES, Rehd.

Picea morindoides, *n. sp.*

Leaves linear, straight and slender, from 2.5 to 3.5 centimetres long, and from 0.7 to 1.2 millimetres broad, acuminate, terminated by a callous sharp tip, somewhat flattened and distinctly keeled on each side (rhomboidal in transverse section), with two white lines on the upper surface consisting each of about six rows of stomata, dark bluish green on the under surface without stomata; furnished beneath the epidermis with one layer of strengthening cells all around, doubled beneath the keel on each side, usually with two resin ducts on the under side close to the epidermis; pulvini slender, pointing forward, not thickened or only slightly thickened at the base. Cones oblong, about 8 centimetres long, 4 centimetres broad when open or scarcely 3 centimetres broad when closed; scales greenish or purplish green, with a purplish border when young, at maturity yellowish brown and lustrous, spatulate-obovate, 2 centimetres long, from 1.2 to 1.4 centimetres broad, the upper part rhombic, with truncate apex and finely serrulate with a slightly wavy margin, appressed when young, spreading upward and rather thin and flexible at maturity; bracts oblong-ovate, acute, from 4 to 5 millimetres long; seeds obovate, about 5 millimetres long, light grayish brown, their oblong or oblong-obovate hyaline wings about 1.2 centimetres in length.

A tree, with whorled spreading branches and slender pendulous or toward the top of the tree spreading branchlets; young branchlets glabrous, yellowish, with pulvini of the same color, becoming yellowish gray the third year; winter-buds ovate, from 5 to 6 millimetres long, obtuse, light yellowish brown, slightly or not at all resinous; bud-scales few, ovate, obtuse.

Native country unknown, probably eastern Asia.

Picea morindoides seems, especially in its cones, most nearly related to *Picea Alcockiana*, Carrière, but differs from that species in its more elongated scales, in its slenderer and longer leaves with stomata only on one side, in the absence of the pubescence on the leading shoots, and in the pendulous branchlets. From the other Piceas with the leaves stomatiferous only on the upper side, like *Picea Omorika*, Bolle, *Picea Sitchensis*, Carrière, *Picea Ajanensis*, Fischer, and *Picea Breweriana*, Watson, it is easily distinguished by its long and slender acuminate leaves, and by the shape of the cone-scales, and from the three first named species also by its habit, which is very similar to that of the Himalayan *Picea Smithiana*, Boissier (*Picea Morinda*, Link); and in allusion to this similarity the specific name *morindoides* is proposed.

The resemblance of the cones to those of *Picea Alcockiana* induced me to consider this tree at first as a variety of that species, an opinion which I communicated to Monsieur Allard; this explains the mention of this name by Mottet in his recently published book on Conifers.[1]

The species is known only in cultivation. The original tree, of uncertain origin, grows in the arboretum of Monsieur G. Allard at Angers, who drew my attention to it when I visited two years ago his interesting and rich collection of trees and shrubs. Young grafted plants propagated from branches sent by Monsieur Allard are now growing at the Arnold Arboretum, but their resistance against the climate of New England has not yet been proved. If this species can support this climate, it will be a valuable substitute for the picturesque *Picea Smithiana*, which is not hardy in northern latitudes.

Alfred Rehder.

Arnold Arboretum.

1 *Picea Alcockiana*, var. *morindoides*, Mottet, *Conifères et Taxacées*, 273 (1902), without description.

EXPLANATION OF THE PLATE.

PLATE XLVIII. PICEA MORINDOIDES.

1. A fruiting branch, natural size.
2. A scale, upper side, with the seeds, enlarged.
3. A scale, lower side, with the bract, enlarged.
4. A seed, enlarged.
5. A leaf, upper side, enlarged.
6. A leaf, under side, enlarged.
7. Cross section of a leaf, enlarged.
8. A winter branch, the leaves removed, natural size.
9. Part of a young branch, enlarged.

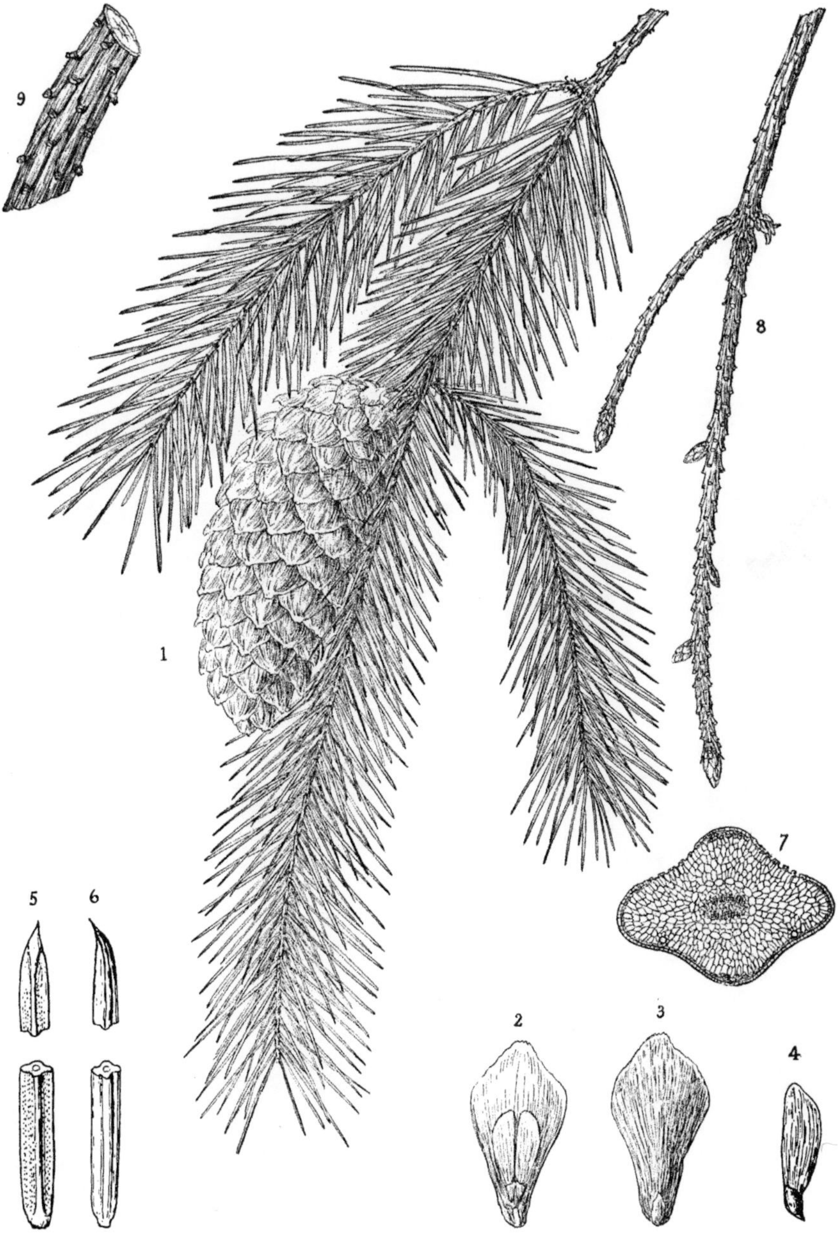

9

1

8

7

5 6

2 3 4

PICEA MORINDOIDES, Rehd.

SOLANUM MOLINUM, Fernald.

Solanum molinum, *n. sp.*

Leaves oblong-lanceolate to oblong, entire, bluntly acuminate or rounded at the apex, acute at the base, from 3 to 9 centimetres long, from 1.3 to 3.5 centimetres broad, dull green and minutely stellate-puberulent above, closely pubescent with whitish stellate hairs beneath; petioles slender, from 7 to 12 millimetres in length. Cymes loosely branched, from 2 to 5 centimetres long, the peduncles and pedicels from 5 to 8 millimetres long, densely canescent-stellate; calyx 4 millimetres high, with lance-ovate lobes; corolla 1.3 centimetres long, the lanceolate lobes pulverulent without; anthers linear, 5 millimetres long, on filaments one fifth as long. Fruit red, smooth, 1 centimetre in diameter.

A shrub, with elongate slender branches covered by a thin translucent early deciduous epidermis, and beneath by a finely furrowed gray-brown thicker bark; branchlets densely canescent-pulverulent, with fine stellate hairs.

Mexico: Oaxaca, near the town of Oaxaca, *C. Conzatti* and *V. González* (No. 1219), June, 1901.

Solanum molinum belongs to the large section Torvaria, and is probably nearest related to *Solanum Hartwegi*, Bentham, a larger species of broad range from the valley of Mexico to the Gulf coast, and south to Costa Rica. From that species *Solanum molinum* is readily distinguished by its smaller leaves and flowers, and by the fine close pubescence, which in *Solanum Hartwegi* is very much longer, and forms a dense lanate covering.

M. L. Fernald.

Gray Herbarium.

EXPLANATION OF THE PLATE.

PLATE XLIX. SOLANUM MOLINUM.

1. A flowering branch, natural size.
2. Vertical section of a flower, enlarged.
3. A stamen, front and rear views, enlarged.
4. A fruiting branch, natural size.
5. A seed, enlarged.

SOLANUM MOLINUM, Fern.

EUPHORBIA LUCIISMITHII, Robins. & Greenm.

Euphorbia Luciismithii, Robins. & Greenm. *Proc. Am. Acad.* xxxii. 36 (1896).

Leaves verticillate, two to five-nate, elliptic or elliptic-oblong, from 1 to 3.5 centimetres long, from 0.5 to 1.5 centimetres broad, obtuse or rounded and subapiculate at the apex, rounded to subacute at the base, glabrous or glabrate above, paler and soft grayish-tomentulose beneath; petioles from 0.5 to 2 centimetres long, tomentulose. Cymes terminal, compound, flat topped, leafy; floral leaves oblanceolate, cuneate, mucronulate, one-nerved, white or rarely red, about 6 millimetres in length, 2.5 millimetres in breadth, involucres campanulate, puberulent, nearly sessile; lobes five, fimbriate glands commonly five, oblong; appendages oblong or subrotund, undulate, white, nearly or quite 1 millimetre long; ovary glabrous. Capsule three-lobed, about 6 millimetres long; seeds oblong, somewhat four-angled, ashy, glabrous, faces rugulose and marked with irregular brown lines.

A tall branching tomentulose glaucous shrub, from 3 to 4.5 metres in height, with subterete striate branches.

Mexico: State of Oaxaca, Rancho de Calderon, altitude 1695 metres, *Rev. Lucius C. Smith* (No. 181), August 13, 1894; Jayacatlan, altitude 1225 metres, *Rev. Lucius C. Smith* (No. 182), September 10 and November 4, 1894; in rocky gulches, Monte Alban, altitude 1785 metres, *C. G. Pringle* (No. 4903), September 14 and November 27, 1894; 9.5 kilometres above Dominguillo, altitude 1385 to 1695 metres, *E. W. Nelson* (No. 1880), October 30, 1894; Oaxaca, altitude 1750 metres, *C. Conzatti* and *V. González* (No. 1043), July and August, 1900.

Euphorbia Luciismithii belongs to the section Alectoroctonum, Boissier, in *De Candolle Prodr.* xv. pt. ii. 59, and is most nearly related to *Euphorbia leucocephala*, Lotsy, *Bot. Gazette*, xx. 350, t. xxiv., from which it differs in the form of the appendages of the glands of the involucre, and in the less conspicuously petaloid bracts of the inflorescence.

J. M. Greenman.

Gray Herbarium.

EXPLANATION OF THE PLATE.

PLATE L. EUPHORBIA LUCIISMITHII.

1. A flowering branch, natural size.
2. Portion of an inflorescence, enlarged.
3. A floral leaf, enlarged.
4. An involucre of staminate flowers, enlarged.
5. An involucre with a pistillate flower, enlarged.
6. An involucre laid open, enlarged.
7. A fimbriated gland of the involucre, enlarged.
8. An appendage of the involucre, enlarged.
9. Part of a fruiting cyme, natural size.
10. A coccus, enlarged.
11. A seed, enlarged.

EUPHORBIA LUCIISMITHII, Robins. & Greenm.

WS - #0028 - 090821 - C0 - 229/152/10 [12] - CB - 9780331834819